STRIVE FOR A 5

Preparing for the AP® Environmental Science Examination

to accompany

FRIEDLAND and RELYEA
ENVIRONMENTAL SCIENCE for the AP® COURSE

FOURTH EDITION

Suzanne Carmody

Kristin Shapiro

 bedford, freeman & worth
publishers

ISBN 978-1-319-53388-5

Bedford, Freeman & Worth Publishers
120 Broadway, New York, NY 10271
bfwpub.com/catalog

Table of Contents

PREFACE

This study guide, *Strive for a 5: Preparing for the AP® Environmental Science Examination*, is designed for use with *Friedland and Relyea's Environmental Science for the AP® Course*, Fourth Edition. It is intended to help you assess your knowledge of the material covered in the textbook, to reinforce key concepts, and to prepare for the AP® Environmental Science Exam. The study guide is organized by unit. As you cover each unit in class, you can use the study guide to check your understanding of the material and practice using new concepts. Each unit contains the following features:

- **Unit Opening Material:** Before you read the unit, preview the key topics, the opening case study, and the "Do the Math" boxes in the unit. Module exercises begin with a list of the learning goals.

- **Key Terms:** The key terms from the chapter are listed in the order they appear in the module. Consider making flashcards for these terms.

- **While You Read:** Questions in this section have been designed to encourage active reading and strong comprehension of the material in the text. Using these questions will improve both your understanding and retention.

- **Practice the Math:** Do these problems to reinforce the important math skills introduced in "Do the Math" in the text.

- **Review Key Terms:** At the end of each module, you can use this exercise to test your knowledge of the new key terms.

- **Check Your Understanding**: At the end of the unit, these questions help solidify key material

- **Practice for Free-Response Questions**: Answer these questions to help you remember and apply unit material and provide writing practice to prepare you for answering free-response questions.

- **Unit Multiple-Choice Review Exams:** Test your understanding and retention at the end of each unit with a practice multiple-choice exam containing 20-30 questions.

When you have finished your work on each unit make sure to take some time to review and reflect using the following steps:

1. What are the challenging concepts from this unit? Identify any concepts you found particularly challenging in this unit. Create a list of topics you need to review in preparation for an exam.
2. What questions do you have about concepts in the unit? Note any further questions you might have about material in the unit. Work with a partner to discuss these questions and ask your teacher for help as needed.
3. Write five possible multiple-choice questions based on this unit. Work with a partner to quiz each other in preparation for an exam.

At the back of the book, you will find two full-length exams with 80 multiple-choice and 3 free-response questions, just like the actual AP® Environmental Science Exam. After completing the exams, check your answers and be sure to review concepts covered in questions you did not answer correctly. Please see your teacher for answers to all questions and practice exams in this guide.

Best of luck with your course this year and on the Exam!

Suzanne Carmody

Kristin Shapiro

MODULE 0

What Is Environmental Science?

Module Summary

This course, and this Module 0, expose the value of environmental science, the field of study that looks at interactions among human systems and those found in nature. This module provides an overview of Environmental Science, explores how humans impact natural systems more profoundly than any other species, and reviews the four big ideas of the course: Energy Transfer, Interactions Between Earth Systems, Interactions Between Different Species and the Environment, and Sustainability. The module also walks through the important uses of and procedures for implementing the scientific method.

Case Study: *The Global Pandemic Was a Global Environmental Science Event Too*

The case study discusses the Covid-19 pandemic which is caused by the SARS coronavirus disease 2019 that spread from humans to animals, as well as from animals to humans. As the human population encroaches into animal habitats, humans have greater contact with animals and possibly more contact with animals that may transmit zoonotic diseases. The case demonstrates how the Covid-19 pandemic impacted birth rates, death rates, and life expectancy resulting in a decreased population growth rate, all of which are important concepts in environmental science. The pandemic also resulted in fewer materials being used, and less fuel consumption, which lead to fewer carbon dioxide emissions. From these few examples, it's clear to see from this module opening case how the pandemic was an impactful global environmental science event.

Do the Math

This module contains the following "Do the Math" box to help prepare you for calculation questions you might encounter on the exam.

- "Range of Electric Vehicles" (page 13)

To make sure you understand the concepts and techniques presented in this box, do the practice problems presented in the text as well as the additional "Practice the Math" problems that appear in this strive guide.

MODULE 0: What Is Environmental Science?

Before You Read the Module

Focus on Learning Goals
Use the module learning goals to guide your reading. On a separate piece of paper, write down each goal and take notes to help you meet each learning goal. After studying this module, you should be able to:
- 0-1 describe the field of environmental science and discuss its importance.
- 0-2 identify ways in which humans have altered and continue to alter our environment.
- 0-3 explain the four "Big Ideas" in environmental science.
- 0-4 describe the scientific method and justify how it is used to design and evaluate information in environmental science.

Key Terms

Environment	Hypothesis	Accuracy
Environmental science	Variable	Precision
Ecosystem	Independent variable	Uncertainty
Biotic	Dependent variable	Inductive reasoning
Abiotic	Null hypothesis	Deductive reasoning
Environmentalism	Control group	Theory
Environmental studies	Natural experiment	First law of thermodynamics
Sustainability	Replication	Second law of
Scientific method	Sample size (n)	thermodynamics

While You Read the Module

Answer the following questions as you read. Use a separate sheet of paper if necessary.

Case Study: The Global Pandemic Was a Global Environmental Science Event Too

1. Describe how Covid-19 might be related to animal interactions that are discussed further in Units 1 and 2.

2. Why might zoonotic diseases increase in the future?

3. Explain how Covid-19 impacted human populations.

4. Describe how carbon dioxide emissions were affected during the Covid-19 pandemic.

5. Give one example of the return to more natural conditions in national parks and other tourist areas during the Covid-19 pandemic.

Environmental science is important if you live on this planet

6. How are humans dependent on Earth?

7. List the environmental conditions that influence life.

8. In what ways do environmental conditions influence humans?

9. Define the study of environmental science.

10. Explain what is meant by the use of the term "system" in "human systems".

11. Give your own example of a human-made system.

12. Give an example of a natural system.

13. Provide an example of an environmental scientist's work from small to large systems.

14. Define ecosystem and the interacting components.

15. Give an example of a biotic and an abiotic component in the environment.

16. Explain how environmental science is different from environmentalism.

17. Figure 0.1: What does the study of environmental science include?

18. Environmental science is a subcategory of what broader field?

Humans impact natural systems in many ways

19. List key biotic and abiotic components in the natural system of a wooded area where a tree has fallen and there has been a fire.

20. Figure 0.2: Identify the different interconnecting systems that operate at different scales for scientists studying the fisheries of the North Atlantic.

21. Identify one example of an intentional environmental change and one example of an unintentional environmental change that human activities have caused.

22. Early humans are thought to have caused the extinction of large mammals. Explain why this happened. Was it intentional or unintentional?

23. How did humans cause changes to the Great Plains? Was this intentional or unintentional?

24. Identify how human well-being has improved over the last two centuries.

25. Identify how human population growth has led to unintentional environmental consequences.

26. Figure 0.3: Describe the differences in the outcomes of the passenger pigeon versus the American bison due to human interaction.

The AP® environmental science course contains four "Big Ideas"

Big Idea 1: Energy Transfer

27. Describe what happens to energy as it flows through systems. What are the consequences of this? How is it significant to environmental science?

Big Idea 2: Interactions Between Earth Systems

28. Systems change over time. Give one example of a change to biogeochemical cycles.

Big Idea 3: Interactions Between Different Species and the Environment

29. Describe how human population growth and technology have impacted Earth.

Big Idea 4: Sustainability

30. Define sustainability.

31. Explain why sustainability is needed for humans worldwide.

The scientific method is an important process in environmental science

The Scientific Method

32. Identify how scientists investigate the natural world.

33. Define scientific method.

34. Figure 0.4: List the steps of the scientific method.

35. Define hypothesis.

36. Visual Representation 0: Describe how burning fossil fuels represents the Big Ideas of Energy Transfer and Interactions Between Earth Systems.

37. Visual Representation 0: How might sustainable farming practices combine the Big Ideas of Sustainability, Interactions between Earth Systems, and Interactions Between Different Species and the Environment?

38. What makes a hypothesis testable? Give an example of a testable hypothesis.

39. Define variable.

40. Explain the differences between independent and dependent variables.

41. Define null hypothesis and give an example.

42. Define control group.

43. Define natural experiment.

44. Define replication.

45. Define sample size.

46. Define accuracy.

47. Define precision.

48. Define uncertainty.

49. Figure 0.5: Explain how the third dartboard demonstrates high accuracy and precision.

50. Contrast inductive and deductive reasoning.

51. Explain an advantage of the dissemination of findings.

52. Define theory.

53. Describe natural law and give an example.

54. Define the first law of thermodynamics.

55. Define the second law of thermodynamics.

After You Read the Module

Review Key Terms
Match the key terms on the left with the definitions on the right.

_____ 1. Environment

_____ 2. Environmental science

_____ 3. Ecosystem

_____ 4. Biotic

_____ 5. Abiotic

_____ 6. Environmentalism

_____ 7. Environmental studies

_____ 8. Sustainability

_____ 9. Scientific method

_____ 10. Hypothesis

_____ 11. Variable

a. A variable that is not dependent on other factors

b. A natural event that acts as an experimental treatment in an ecosystem.

c. The field of study that looks at interactions among human systems and those found in nature.

d. In a scientific investigation, a group that experiences exactly the same conditions as the experimental group, except for the single variable under study.

e. How close a measured value is to the actual or true value.

f. The sum of all the conditions surrounding us that influence life.

g. The number of times a measurement is replicated in data collection.

h. Any categories, conditions, factors, or traits that differ in the natural world or in experimental situations.

i. An estimate of how much a measured or calculated value differs from a true value.

j. A testable conjecture about how something works.

k. Living.

_____ 12. Independent variable

l. A prediction that there is no difference between the groups or conditions that are being compared.

_____ 13. Dependent variable

m. A particular location on Earth with interacting biotic and abiotic components.

_____ 14. Null hypothesis

n. A theory with no known exception that states that energy is neither created nor destroyed but it can change from one form to another.

_____ 15. Natural experiment

o. The data collection procedure of taking repeated measurements.

_____ 16. Control group

p. The process of making general statements from specific facts or examples.

_____ 17. Replication

q. Nonliving.

_____ 18. Sample size (*n*)

r. An objective method to explore the natural world, draw inferences from it, and predict the outcome of certain events, processes, or changes.

_____ 19. Accuracy

s. Using Earth's resources in a way that does not jeopardize future generations from engaging in similar activities.

_____ 20. Precision

t. A hypothesis that has been repeatedly tested and confirmed by multiple groups of researchers and has reached wide acceptance.

_____ 21. Uncertainty

u. The field of study that includes environmental science and additional subjects such as environmental policy, economics, literature, and ethics.

_____ 22. Inductive reasoning

v. A theory with no known exception that states that when energy is transformed, the quantity of energy remains the same, but its ability to do work diminishes.

_____ 23. Deductive reasoning

w. A variable that is dependent on other factors.

_____ 24. Theory

x. A social movement that seeks to protect the environment through lobbying, activism, and education.

_____ 25. First law of thermodynamics

y. The process of applying a general statement to specific facts or situations.

_____ 26. Second law of thermodynamics

z. How close the repeated measurements of a sample are to one another.

MODULE 0 Review Exercises

✓ Check Your Understanding

Review "Learning Goals Revisited" in your textbook. Compare the notes you took while reading the module. Complete these exercises to review the module. Use a separate sheet of paper if necessary.

1. Come up with a hypothetical experiment you plan on conducting.

    (a) Describe what a hypothesis is and provide the hypothesis for your proposed investigation.

    (b) Define independent and dependent variable. Identify the independent and dependent variable in your proposed investigation.

    (c) Describe a possible control group for your investigation and defend its use as a control.

    (d) Identify a null hypothesis for your proposed investigation.

    (e) Explain how you would collect data for your proposed investigation that is both precise and accurate.

    (f) Defend the need for multiple trials in your proposed experiment.

2. Explain how environmental science is part of a larger environmental studies discipline.

3. Review the Big Ideas of Environmental Science.

    (a) Explain Big Idea 1: Energy Transfer and provide one example you are familiar with.

    (b) Explain Big Idea 2: Interactions Between Earth Systems and provide one example you are familiar with.

(c) Explain Big Idea 3: Interactions Between Different Species and the Environment and provide one example you are familiar with.

(d) Explain Big Idea 4: Sustainability and provide one example you are familiar with.

Practice for Free-Response Questions

Complete this exercise to build and practice the skills you will need to answer free-response questions on the exam. Use a separate sheet of paper if necessary.

1. Explain why when humans interact with different species and the environment they should always have sustainability in mind.

2. A student in an AP® Environmental Science class shares a hypothesis about cars releasing particulate matter. Another student claims that because there are decades of research to support this statement, it isn't a hypothesis but instead a theory. Provide a justification for why student two is correct.

UNIT 1

The Living World: Ecosystems

Unit Summary

Unit 1 of this course lays the foundation for understanding how Earth works as a system and the relationships between the abiotic and biotic components. The beginning of this unit focuses on biotic factors such as species interactions: predation, herbivory, resource partitioning, and symbiosis. The rest of Unit 1 focuses on abiotic factors such as how terrestrial and aquatic biomes are formed based on their location on Earth. Modules 4 and 5 focus on how nutrients such as carbon, phosphorus, and nitrogen are continually cycled between Earth's systems. The unit concludes with the cycling of energy and matter in Modules 6 and 7. The unit consists of seven modules.

MODULES IN THIS UNIT

Module 1: Introduction to Ecosystems
Module 2: Terrestrial Biomes
Module 3: Aquatic Biomes
Module 4: The Carbon and Nitrogen Cycles
Module 5: The Phosphorus and Hydrologic (Water) Cycles
Module 6: Primary Productivity
Module 7: Trophic Levels, Energy Flow and the 10% Rule, Food Chains, and Food Webs

---

Unit Opening Case: *Growing Grapes to Make Fine Wine*
This unit opening case illustrates that different regions of the world contain distinct climates and that these climates affect the species that can live in each region. Students learn that certain regions have comparable climates and therefore support similar plant and animal communities. The story of wine grapes further demonstrates that as climate changes, we can expect changes in the species that live in these regions as well as changes in the way humans use these ecosystems.

Do the Math
This unit contains the following "Do the Math" boxes to help prepare you for calculation questions you might encounter on the exam.
- "Organic Matter Inputs Among Terrestrial Biomes" (page 38)
- "Measuring Inputs of Nitrogen and Phosphorus in an Urban Environment" (page 62)
- "Raising Mangoes" (page 64)
- "Calculating NPP, GPP, and R" (page 69)

To make sure you understand the concepts and techniques presented in this box, do the practice problems presented in the text as well as the additional "Practice the Math" problems that appear in Modules 2, 5, and 6 of this study guide.

MODULE 1: Introduction to Ecosystems

Before You Read the Module

Focus on Learning Goals
Use the module learning goals to guide your reading. On a separate piece of paper, write down each goal and take notes to help you meet each learning goal. After studying this module, you should be able to:
- 1-1 explain how we define ecosystem boundaries.
- 1-2 describe how competing species respond to limited resources.
- 1-3 identify which species interactions involve one species consuming another species.
- 1-4 describe which species interactions cause neutral or positive effects on both species.
- 1-5 explain how invasive species represent novel species interactions.

Key Terms

Climate	Resource partitioning	Commensalism
Weather	Predation	Native species
Community ecology	Parasitoid	Exotic species (Alien
Symbiosis	Parasitism	species)
Biosphere	Pathogen	Invasive species
Competition	Herbivory	
Competitive exclusion	Mutualism	
principle	Photosynthesis	

While You Read the Module
Answer the following questions as you read. Use a separate sheet of paper if necessary.

Case Study: Growing Grapes to Make Fine Wines

1. What regions of the world are best known for fine winemaking?

2. What are the best growing conditions for grapes?

3. What do the five best regions for growing grapes have in common?

4. What is expected to happen in the future to California vineyards and why?

5. Describe what changes have occurred in French and English vineyards.

6. Define climate.

7. Define weather.

Module 1: Introduction to Ecosystems

8. Define community ecology.

9. Define symbiosis.

Ecosystem boundaries are defined by biotic and abiotic components or defined by humans

10. What are the characteristics of a given ecosystem dependent upon?

11. Describe two examples of ecosystems that have well-defined boundaries.

12. Explain why ecosystem boundaries are often subjective and give examples.

13. Figure 1.2a: Why does the ecosystem boundary of the Greater Yellowstone Ecosystem include national parks, national forests, and private land?

14. Define biosphere.

Competition for limited resources between species can lead to resource partitioning

Competition

15. Define competition.

16. Define competitive exclusion principle.

17. Figure 1.3: Explain how the different Paramecium species in the experiment affected each other's population growth.

Resource Partitioning

    18. Define resource partitioning.

    19. Figure 1.4: Describe the example of resource partitioning.

    20. Explain how plants use resource partitioning.

    21. Explain Darwin's finches and their morphological resource partitioning.

    22. How does resource partitioning help competition among species?

Interactions involving one species consuming another include predation, parasitism, and herbivory

Predation

    23. Define predation and give an example.

    24. Define parasitoids and give an example.

    25. List the different types of defenses used to avoid predators.

Parasitism

    26. Define parasitism and give an example.

    27. Define a pathogen and list three examples.

Herbivory

    28. Define herbivory.

29. List the examples of well-known herbivores, for both terrestrial and water ecosystems.

30. Figure 1.8: How can herbivory have a detrimental effect?

Species interactions that cause neutral or positive effects include mutualisms and commensalisms

Mutualism

31. Why is the relationship between plants and their pollinators one of the most ecologically important?

32. Define mutualism.

33. Figure 1.9: Explain the well-studied plant-animal interaction of the acacia trees and *Pseudomyrmex* ants in Central America.

34. List two other examples of mutualism.

35. Algae is said to provide sugars for the coral via photosynthesis. Describe the process of photosynthesis.

Commensalisms

36. Define commensalism and provide an example.

37. Table 1.1: Copy Table 1.1 into your notes, explain the meaning of (+), (-), and (0).

Invasive species represent novel species interactions due to a lack of evolutionary history

38. Define a native species.

39. Define an exotic species (alien species) and give an example.

40. Explain how rats moved around the world and their detrimental environmental impact.

41. Define an invasive species.

42. Figure 1.11: Why was the kudzu vine introduced to the United States in 1876?

43. Figure 1.11 Explain how the kudzu vine became an invasive species.

After You Read the Module

Review Key Terms

Match the key terms on the left with the definitions on the right.

_____ 1. Climate	a. When two species evolve to divide a resource based on differences in their behavior or morphology.
_____ 2. Weather	b. An interaction in which one organism lives on or in another organism, referred to as the host.
_____ 3. Community ecology	c. The process by which plants and algae use solar energy to convert carbon dioxide (CO_2) and water (H_2O) into glucose ($C_6H_{12}O_6$) and oxygen (O_2).
_____ 4. Symbiosis	d. An interaction in which an animal consumes plants or algae.
_____ 5. Biosphere	e. The struggle of individuals, either within or between species, to obtain a shared limiting resource.
_____ 6. Competition	f. The principle stating that two species competing for the same limiting resource cannot coexist.
_____ 7. Competitive exclusion principle	g. An interaction between two species that increases the chances of survival or reproduction for both species.
_____ 8. Resource partitioning	h. An interaction between two species in which one species benefits and the other species is neither harmed nor helped.
_____ 9. Predation	i. The average weather that occurs in a given region over a long period of time.
_____ 10. Parasitoid	j. A species living outside its historical range.
_____ 11. Parasitism	k. The study of interactions among species.
_____ 12. Pathogen	l. A specialized type of predator that lays eggs inside other organisms — referred to as its host.
_____ 13. Herbivory	m. The region of our planet where life resides.
_____ 14. Mutualism	n. A species that lives in its historical range, typically where it has lived for thousands or millions of years.
_____ 15. Photosynthesis	o. The short-term conditions of the atmosphere in a local area that include temperature, humidity, clouds, precipitation, and wind speed.
_____ 16. Commensalism	p. A species that spreads rapidly across large areas and causes harm.
_____ 17. Native species	q. A parasite that causes disease in its host.
_____ 18. Exotic species (Alien species)	r. Two species living in a close and long-term association with one another in an ecosystem.
_____ 19. Invasive species	s. An interaction in which one animal typically kills and consumes another animal.

MODULE 2: Terrestrial Biomes

Before You Read the Module

Focus on Learning Goals

Use the module learning goals to guide your reading. On a separate piece of paper, write down each goal and take notes to help you meet each learning goal. After studying this module, you should be able to:

- 2-1 explain how we define terrestrial biomes.
- 2-2 describe the information contained in climate diagrams.
- 2-3 identify the nine terrestrial biomes.
- 2-4 describe the causes of changing boundaries of terrestrial biomes.

Key Terms

Invasive species	Permafrost	Temperate grassland (Cold
Biome	Taiga (Boreal forest)	desert)
Terrestrial biome	Temperate rainforest	Tropical rainforest
Aquatic biome	Temperate seasonal forest	Savanna (Tropical seasonal
Habitat	Shrubland (Woodland)	forest)
Tundra		Hot desert

While You Read the Module

Answer the following questions as you read. Use a separate sheet of paper if necessary.

Module 2 Terrestrial Biomes

1. Define biome

2. What two factors determine the survival of specific species in a biome?

Terrestrial biomes are defined by the dominant plant growth forms

3. Figure 2.1: Compare and contrast the photographs.

4. Define terrestrial biomes.

5. Define aquatic biomes.

6. Figure 2.2: What two characteristics determine where biomes are located on the graph?

7. Define habitat.

8. Compare a habitat to a biome.

9. Figure 2.3: Look at the map and identify the biome in which you live.

Climate diagrams illustrate patterns of annual temperature and precipitation

10. How does a climate diagram illustrate patterns of annual temperature and precipitation?

11. How can climate diagrams be helpful?

12. Figure 2.4: Identify what each color represents on the climate diagrams.

13. Figure 2.4: Explain what the location of the precipitation and temperature lines on a climate diagram tell us.

Terrestrial biomes range from tundra to tropical forests to deserts

Tundra

14. Define tundra.

15. List the locations and characteristics of the tundra.

16. Define permafrost.

17. Identify how humans have affected this biome.

Taiga

18. Define taiga. Is it different from the boreal forest?

19. List the location and characteristics of the taiga.

20. Identify how humans have affected this biome.

Temperate Rainforest

21. Define temperate rainforest.

22. List the location and characteristics of the temperate rainforest.

23. Identify how humans have impacted this biome.

Temperate Seasonal Forest

24. Define temperate seasonal forest.

25. List the location and characteristics of the temperate seasonal forest.

26. Identify how humans have affected this biome.

Shrubland

27. Define shrubland. Is it different from a woodland?

28. List the location and characteristics of the shrubland.

29. Identify how humans have impacted this biome.

Temperate Grassland

30. Define temperate grassland. Is it different from a cold desert?

31. List the location and characteristics of the temperate grassland.

32. Identify how humans have affected this biome.

Tropical Rainforest

33. Define tropical rainforest.

34. List the location and characteristics of the tropical rainforest.

35. Identify how humans have impacted this biome.

Savanna

36. Define savanna. Is it different from a tropical seasonal forest?

37. List the location and characteristics of the savanna.

38. Identify how humans have affected this biome.

Hot Desert

39. Define hot desert. Is it different from a subtropical desert?

40. List the locations and characteristics of the hot desert.

41. Identify how humans have affected this biome.

Biome boundaries shift as climates change

42. Why do biome boundaries shift during ice ages?

Practice the Math: Organic Matter Inputs Among Terrestrial Biomes

Read "Do the Math: Organic Matter Inputs Among Terrestrial Biomes" on page 38. Try the "Your Turn." For more math practice, do the following problem. Remember to show your work. Use a separate sheet of paper if necessary.

Four additional biomes with five sites each were added to the previous sample. Scientists set out containers in different biomes to determine how much dead plant material falls to the ground.

Biome	Sample 1 (tons/ha/year)	Sample 2 (tons/ha/year)	Sample 3 (tons/ha/year)	Sample 4 (tons/ha/year)	Sample 5 (tons/ha/year)
Chaparral	5.3	2.7	3.9	4.0	3.6
Subtropical Desert	0.7	1.2	0.2	1.0	0.4
Temperate Rainforest	17.3	18.2	13.7	12.9	14.4
Savanna	6.2	5.4	7.6	5.9	5.4

(a) Calculate the average amount of dead plant material that fell to the ground in each of the four biomes.

(b) Convert each biome's average from tons/ha/year to $kg/m^2/year$.

Review Key Terms

Match the key terms on the left with the definitions on the right.

_____	1. Biome	a. An area where a particular species lives in nature.
_____	2. Terrestrial Biome	b. A biome characterized by hot, dry summers and mild, rainy winters.
_____	3. Aquatic biome	c. An impermeable, permanently frozen layer of soil.
_____	4. Habitat	d. A warm and wet biome found between 20° N and 20° S of the equator, with little seasonal temperature variation and high precipitation.
_____	5. Tundra	e. A forest biome made up primarily of coniferous evergreen trees that can tolerate cold winters and short growing seasons.
_____	6. Permafrost	f. A geographic region of land categorized by a particular combination of average annual temperature, annual precipitation, and distinctive plant growth forms.
_____	7. Taiga (Boreal forest)	g. A biome characterized by cold, harsh winters, and hot, dry summers.
_____	8. Temperate rainforest	h. A coastal biome typified by moderate temperatures and high precipitation.
_____	9. Temperate seasonal forest	i. The plants and animals that are found in a particular region of the world.
_____	10. Shrubland (Woodland)	j. A biome marked by warm temperatures and distinct wet and dry seasons.
_____	11. Temperate grassland (Cold desert)	k. A cold and treeless biome with low-growing vegetation.
_____	12. Tropical rainforest	l. An aquatic region characterized by a particular combination of salinity, depth, and water flow.
_____	13. Savanna (Tropical seasonal forest)	m. A biome located at roughly 30° N and 30° S, and characterized by hot temperatures, extremely dry conditions, and sparse vegetation.
_____	14. Hot desert	n. A biome with warm summers and cold winters with over 1 m (39 inches) of annual precipitation.

MODULE 3: Aquatic Biomes

Before You Read the Module

Focus on Learning Goals

Use the module learning goals to guide your reading. On a separate piece of paper, write down each goal and take notes to help you meet each learning goal. After studying this module, you should be able to:

- 3-1 identify the major freshwater biomes.
- 3-2 identify the major marine biomes.

Key Terms

Freshwater biomes	Mesotrophic	Coral reef
Littoral zone	Eutrophic	Coral bleaching
Limnetic zone	Freshwater wetland	Open ocean
Phytoplankton	Estuary	Photic zone
Profundal zone	Salt marsh	Aphotic zone
Benthic zone	Mangrove swamp	Chemosynthesis
Oligotrophic	Intertidal zone	

While You Read the Module

Answer the following questions as you read. Use a separate sheet of paper if necessary.

Module 3: Aquatic Biomes

1. What physical characteristics categorize aquatic biomes?

2. What are the two categories of aquatic biomes? Give examples of each.

3. What are the major determinants of plant and animal species found in aquatic biomes?

4. Figure 3.1: List the five repositories of Earth's water.

Freshwater biomes have low salinity

5. What are characteristics of freshwater biomes? What aquatic biomes are included?

Streams and Rivers

6. What characterizes a stream?

7. What characterizes a river?

8. List the characteristics of a fast-moving body of water.

9. List the characteristics of a slower-moving body of water.

10. Identify how humans have affected stream and rivers.

Lakes and Ponds

11. What determines whether a body of water is a lake or a pond?

12. Define littoral zone.

13. Define limnetic zone.

14. Define phytoplankton.

15. Define profundal zone.

16. Define benthic zone.

17. How are lakes classified?

18. Define oligotrophic.

19. Define mesotrophic.

20. Define eutrophic.

21. Identify how humans have affected lakes and ponds.

Freshwater Wetlands

22. Define freshwater wetlands.

23. List the other names of freshwater wetlands.

24. Describe the ecosystem services freshwater wetlands provide.

25. Figure 3.5: Identify the unique vegetation in each of the photos.

26. Identify how humans have affected freshwater wetlands.

Marine biomes have high salinity

Estuaries and Salt Marshes

27. What are characteristics of marine biomes? What aquatic biomes are included?

28. Define estuaries.

29. Define salt marshes.

30. Describe the importance of the habitat an estuary or salt water marsh provides.

31. Identify how humans have affected salt marshes.

Mangrove Swamps

    32. Define mangrove swamps.

    33. What is unique about mangrove trees?

    34. Describe the benefits of mangrove swamps.

    35. Identify how humans have affected mangrove swamps.

Intertidal Zones

    36. Define intertidal zones.

    37. Describe the harsh conditions that organisms are exposed to during low tide.

    38. Identify how humans have affected intertidal zones.

    39. Figure 3.8: Distinguish between the amount of biodiversity found in the splash zone and that found in the mid-tide zone.

Coral Reefs

    40. Define coral reefs.

    41. Describe corals.

    42. Describe the relationship between coral and single-celled algae.

    43. Describe how coral reefs become massive.

44. Explain the diversity of a coral reef.

45. Define coral bleaching.

46. Identify how humans have affected coral reefs and the challenges coral reefs face.

47. Figure 3.9: Describe the differences between the two photographs.

The Open Ocean

48. Define open ocean.

49. Figure 3.10: Describe where each of the zones of the open ocean are found.

50. What characterizes the photic zone? What important interaction with the Earth Systems occurs here?

51. What characterizes the aphotic zone?

52. Describe the benthic zone.

53. What forms the base of the food webs in the open ocean?

54. Define chemosynthesis.

55. Where does chemosynthesis take place?

56. Where do the nutrients for the tube worms come from?

Review Key Terms

Match the key terms on the left with the definitions on the right.

_____	1.	Freshwater biomes	a.	An aquatic biome that is submerged or saturated by water for at least part of each year, but shallow enough to support emergent vegetation.
_____	2.	Littoral zone	b.	Floating algae.
_____	3.	Limnetic zone	c.	The narrow band of coastline that exists between the levels of high tide and low tide.
_____	4.	Phytoplankton	d.	An area along the coast where the fresh water of rivers mixes with salt water from the ocean.
_____	5.	Profundal zone	e.	A phenomenon in which algae inside corals die, causing the corals to turn white.
_____	6.	Benthic zone	f.	Found along the coast in temperate climates, a marsh containing nonwoody emergent vegetation.
_____	7.	Oligotrophic	g.	Describes a lake with a low level of phytoplankton due to low amounts of nutrients in the water.
_____	8.	Mesotrophic	h.	A zone of open water in lakes and ponds as deep as the sunlight can penetrate.
_____	9.	Eutrophic	i.	The muddy bottom of a lake, pond, or ocean beneath the limnetic and profundal zones.
_____	10.	Freshwater wetland	j.	A process used by some bacteria to generate energy with methane and hydrogen sulfide.
_____	11.	Estuary	k.	A region of water where sunlight does not reach, below the limnetic zone in very deep lakes.
_____	12.	Salt marsh	l.	A swamp that occurs along tropical and subtropical coasts, and contains salt-tolerant trees with roots submerged in water.
_____	13.	Mangrove swamp	m.	Represents Earth's most diverse marine biome, and are found in warm, shallow waters beyond the shoreline in tropical regions.
_____	14.	Intertidal zone	n.	Deep-ocean water, located away from the shoreline where sunlight can no longer reach the ocean bottom.
_____	15.	Coral reef	o.	Describes a lake with a moderate level of fertility.
_____	16.	Coral bleaching	p.	The upper layer of ocean water in the ocean that receives enough sunlight for photosynthesis.
_____	17.	Open ocean	q.	Categorized as streams and rivers, lakes and ponds, or freshwater wetlands.
_____	18.	Photic zone	r.	The deeper layer of ocean water that lacks sufficient sunlight for photosynthesis.
_____	19.	Aphotic zone	s.	Describes a lake with a high level of fertility.
_____	20.	Chemosynthesis	t.	The shallow zone of soil and water in lakes and ponds near the shore where most algae and emergent plants such as cattails grow.

MODULE 4: The Carbon and Nitrogen Cycles

Before You Read the Module

Focus on Learning Goals

Use the module learning goals to guide your reading. On a separate piece of paper, write down each goal and take notes to help you meet each learning goal. After studying this module, you should be able to:

- 4-1 explain how carbon cycles within ecosystems.
- 4-2 describe how nitrogen cycles within ecosystems.

Key Terms

Biogeochemical cycle	Global warming	Mineralization
Reservoirs	Limiting nutrient	(Ammonification)
Carbon cycle	Nitrogen cycle	Denitrification
Aerobic respiration	Nitrogen fixation	Anaerobic
Steady state	Nitrification	Aerobic
Greenhouse gases	Assimilation	Leaching

While You Read the Module

Answer the following questions as you read. Use a separate sheet of paper if necessary.

Module 4: The Carbon and Nitrogen Cycles

1. Why is the fact that ecosystems are a closed system for matter significant for nutrient cycling?

2. Define biogeochemical cycles.

3. Define a biogeochemical reservoir. In this context, what is a "source" and what is a "sink"?

The carbon cycle moves carbon between air, water, and land

4. Define carbon cycle.

The Carbon Cycle

5. Distinguish between the fast and slow processes that drive the carbon cycle.

6. Define photosynthesis.

7. Describe aerobic respiration.

8. Describe how carbon is returned to the water or air after an organism dies.

9. Figure 4.1: List the seven processes that drive the carbon cycle.

10. Figure 4.1: How does the exchange of carbon occur between the atmosphere and the oceans?

11. How does CO_2 become part of the sedimentation and burial process?

12. How does organic carbon become incorporated into sediments?

13. What is the "steady sate" part of the carbon cycle?

14. Explain the difference between how extraction and combustion of fossil fuels each affect the carbon cycle.

Human Impacts on the Carbon Cycle

15. Under what conditions is the exchange of carbon between Earth's surface and atmosphere in a steady state?

16. Define greenhouse gases.

17. What keeps Earth's temperature so constant?

18. Figure 4.2: Notice the solar energy entering the heat-trapping greenhouse gases. Identify and describe how the heat moves in the diagram.

19. Identify and explain two ways humans have had an influence in carbon cycling.

20. How have carbon dioxide concentrations changed since 1600?

21. Figure 4.3: Describe the correlation of the carbon dioxide levels with the global temperatures shown on the graph.

22. What do scientists believe is the cause in the rise of atmospheric carbon dioxide?

23. Define global warming.

The nitrogen cycle includes many chemical transformations

24. What two things is nitrogen used for? What are the functions of each?

25. Define limiting nutrient.

26. What does the lack of a limiting nutrient lead to?

The Nitrogen Cycle

27. Define the nitrogen cycle.

28. Figure 4.4: List the five major transformations in the nitrogen cycle.

29. How is the nitrogen cycle different from the carbon cycle?

30. How much of Earth's atmosphere is nitrogen?

31. Define nitrogen fixation.

32. What is the biotic process for nitrogen fixation?

33. Identify the two types of abiotic processes for nitrogen fixation.

34. Define nitrification.

35. What is the difference between nitrite and nitrate?

36. Define assimilation.

37. Define mineralization.

38. Define ammonification.

39. Define denitrification.

40. Table 4.1: Write down the molecular formula for each transformation in the nitrogen cycle, differentiating between biotic and abiotic.

41. What are aerobic conditions?

Human Impacts on the Nitrogen Cycle

42. Where is nitrogen a limiting nutrient?

43. Define leaching.

44. How have humans impacted the nitrogen cycle?

45. What effects on the environment have been observed by adding nitrogen to the soils?

Review Key Terms
Match the key terms on the left with the definitions on the right.

_____ 1. Biogeochemical cycle

_____ 2. Reservoirs

_____ 3. Carbon cycle

_____ 4. Aerobic respiration

_____ 5. Steady state

_____ 6. Greenhouse gases

_____ 7. Global warming

_____ 8. Limiting nutrient

_____ 9. Nitrogen cycle

_____ 10. Nitrogen fixation

_____ 11. Nitrification

_____ 12. Assimilation

_____ 13. Mineralization (Ammonification)

_____ 14. Denitrification

_____ 15. Anaerobic

_____ 16. Aerobic

_____ 17. Leaching

a. A process by which plants and algae incorporate nitrogen into their tissues.

b. The movements of matter within and between ecosystems involving cycles of biological, geological, and chemical processes.

c. Gases in Earth's atmosphere that trap heat near the surface.

d. The process that converts nitrogen gas in the atmosphere (N_2) into forms of nitrogen that plants and algae can use.

e. The conversion of ammonia (NH_4^+) into nitrite (NO_2^-) and then into nitrate (NO_3^-).

f. The process by which fungal and bacterial decomposers break down the organic matter found in dead bodies and waste products and convert these organic compounds back into inorganic compounds, such as inorganic ammonium (NH_4^+).

g. A nutrient required for the growth of an organism but available in a lower quantity than other nutrients.

h. The increase in global temperatures due to humans producing more greenhouse gases.

i. The conversion of nitrate (NO_3^-) in a series of steps into the gases nitrous oxide (N_2O) and, eventually, nitrogen gas (N_2), which is emitted into the atmosphere.

j. When a system's inputs equal outputs, so that the system is not changing over time.

k. The movement of carbon around the biosphere among reservoir sources and sinks.

l. The process by which cells convert glucose and oxygen into energy, carbon dioxide, and water.

m. The components of the biogeochemical cycle that contain the matter, including air, water, and organisms.

n. An environment that lacks oxygen.

o. An environment with abundant oxygen.

p. A process in which dissolved molecules are transported through the soil via groundwater.

q. The movement of nitrogen around the biosphere among reservoir sources and sinks.

MODULE 5: The Phosphorus and Hydrologic (Water) Cycles

Before You Read the Module

Focus on Learning Goals
Use the module learning goals to guide your reading. On a separate piece of paper, write down each goal and take notes to help you meet each learning goal. After studying this module, you should be able to:
- 5-1 explain how phosphorus cycles within ecosystems.
- 5-2 describe how water cycles within ecosystems.

Key Terms

Phosphorus cycle	Dead zone	Evapotranspiration
Algal bloom	Hydrologic cycle	Runoff
Hypoxic	Transpiration	

While You Read the Module

Answer the following questions as you read. Use a separate sheet of paper if necessary.

Module 5: The Phosphorus and Hydrologic (Water) Cycles

The phosphorus cycle moves between land and water

1. Why do organisms need phosphorus?

The Phosphorus Cycle

2. Define the phosphorus cycle.

3. Where does the phosphorus cycle operate?

4. What is the form of phosphorus found in nature?

5. Describe the biotic and abiotic processes that affect the phosphorus cycle.

6. Describe the geologic forces of the phosphorus cycle.

7. Where is phosphorus a limiting nutrient?

Human Impacts on the Phosphorus Cycle

8. Describe how humans have affected the phosphorus cycle by using fertilizers.

9. Define algal bloom.

10. Define hypoxic.

11. Define dead zone.

12. Describe how human use of detergents has affected the phosphorus cycle.

13. Describe how agricultural expansion and phosphorus concentrations hurt wetlands, such as the Florida Everglades?

The hydrologic cycle moves water through the biosphere

14. What is the importance of water?

15. Define the hydrologic cycle.

The Hydrologic Cycle

16. Define transpiration.

17. What are the three distinct routes water takes when it falls on land?

18. Define evapotranspiration.

19. Define runoff.

20. Figure 5.4: Describe infiltration.

21. Figure 5.4: Analyze the diagram: What are the driving mechanisms for the hydrologic cycle? Explain.

Human Activities and the Hydrologic Cycle

22. List the human activities that affect the hydrologic cycle and the result of each.

Practice the Math: Measuring Inputs of Nitrogen and Phosphorus in an Urban Environment

Read "Do the Math: Measuring Inputs of Nitrogen and Phosphorus in an Urban Environment" on page 62. Try the "Your Turn." For more math practice, do the following problem. Remember to show your work. Use a separate sheet of paper if necessary.

Cities and rural areas have different sources of nitrogen and phosphorus due to the differing activities and industries that occur in each. Using the data in the chart, calculate the percentage each source contributes to the total nitrogen and total phosphorus additions to the river.

Source	Nitrogen (kg/km^2/year)	Phosphorus (kg/km^2/year)
Atmospheric deposition	1,300	40
Agricultural fertilizer	7,400	20
Agricultural pesticides	0	30
Septic tank seepage	100	10
Ground-water discharge	200	10

Practice the Math: Raising Mangoes

Read "Do the Math: Raising Mangoes" on page 64. Try the "Your Turn." For more math practice, do the following problems. Remember to show your work. Use a separate piece of paper if necessary.

Banana saplings cost $8 each. Once the trees mature, each tree will produce $50 worth of fruit per year. A village of 300 people decides to pool its resources to establish a community banana plantation. Their goal is to generate an income of $200 per year for each person in the village.

(a) How many mature trees will the village need to meet the goal?

(b) Each tree requires 10 m^2 of space. How many hectares must the village set aside for the plantation? (1 m^2 = 0.0001 ha)

(c) Each tree requires 15 L of water to be pumped in every day during the 6 hot months of the year (180 days) and no additional water during the other 6 months. How many liters of water are needed each year?

After You Read the Module

Review Key Terms
Match the key terms on the left with the definitions on the right.

_____ 1. Phosphorus cycle	a. The movement of phosphorus around the biosphere among reservoir sources and sinks.
_____ 2. Algal bloom	b. The movement of water around the biosphere among reservoir sources and sinks.
_____ 3. Hypoxic	c. The release of water from leaves into the atmosphere during photosynthesis.
_____ 4. Dead zone	d. Low in oxygen.
_____ 5. Hydrologic cycle	e. Water that moves across the land surface and into streams and rivers.
_____ 6. Transpiration	f. When oxygen concentrations become so low that it kills fish and other aquatic animals.
_____ 7. Evapotranspiration	g. A rapid increase in the algal population of a waterway.
_____ 8. Runoff	h. The combined amount of evaporation and transpiration.

MODULE 6: Primary Productivity

Before You Read the Module

Focus on Learning Goals

Use the module learning goals to guide your reading. On a separate piece of paper, write down each goal and take notes to help you meet each learning goal. After studying this module, you should be able to:

- 6-1 describe how photosynthesis and respiration affect energy flow.
- 6-2 calculate gross primary productivity and net primary productivity.
- 6-3 explain why primary productivity has a low efficiency.
- 6-4 explain why some ecosystems are much more productive than others.

Key Terms

Producers (Autotrophs)
Cellular respiration
Anaerobic respiration
Primary productivity

Gross primary productivity
(GPP)
Net primary productivity
(NPP)

Biomass
Standing crop

While You Read the Module

Answer the following questions as you read. Use a separate sheet of paper if necessary.

Module 6: Primary Productivity

Photosynthesis captures energy while respiration releases energy

Photosynthesis

1. Define producers, or autotrophs.

2. Define photosynthesis.

3. Figure 6.1: Write the equation for photosynthesis.

4. Figure 6.1: What organisms perform photosynthesis?

Cellular Respiration

5. Define cellular respiration.

6. Define aerobic respiration.

7. Define anaerobic respiration.

8. Explain how producers carry out aerobic respiration and photosynthesis and the net effect.

Primary productivity is the rate of converting solar energy into organic compounds over time

Gross versus Net Primary Productivity

9. What does the amount of energy in an ecosystem determine?

10. Define gross primary productivity (GPP).

11. Define net primary productivity (NPP).

12. Show the equation for net primary productivity (NPP).

13. Explain gross primary productivity (GPP) and net primary productivity (NPP) in terms of a paycheck.

Measuring Gross versus Net Primary Productivity

14. How can scientists measure CO_2 movement during photosynthesis? Write the equation.

15. Figure 6.2: How can scientists derive gross primary productivity?

16. Define biomass.

17. Define standing crop.

18. Explain the difference between standing crop and productivity. Give an example.

Primary productivity is not an efficient process

19. Figure 6.3: Figure 6.3 is a model of gross and net primary productivity. Create your own model of gross and net primary productivity, using colors or symbols.

20. Figure 6.3: Why is converting sunlight into chemical energy an inefficient process?

21. Figure 6.3: Where does the ninety-nine percent of unused sunlight go?

22. Of the one percent of the Sun's energy that is captured, what percent is used to support the producer's growth and reproduction?

23. Figure 6.4: Describe what happens to the various wavelengths of light as they penetrate the depths of water. How is this similar to white light passing through a glass prism?

Some ecosystems are much more productive than others

24. What factors determine where producers grow best?

25. After an ecosystem changes, what can you learn from the amount of stored energy (NPP)?

26. Figure 6.5: Identify the ecosystems with the greatest amount of net primary productivity and the ecosystems with the lowest amount of net primary productivity. What factors contribute to these rankings?

Read "Do the Math: Calculating NPP, GPP, and R" on page 69. Try the "Your Turn." For more math practice, do the following problems. Remember to show your work. Use a separate piece of paper if necessary.

(a) If the open ocean has a net primary productivity of 200 g C/m²/year, and a gross primary productivity of 320 g C/ m²/year what is the respiration (R)?

(b) Most temperate seasonal forests have now been converted to agricultural lands. Suppose they have a net primary productivity of 1,200 g C/m²/year, and a respiration of 320 g C/m²/year, what is the gross primary productivity?

(c) Calculate the percent change in net primary productivity that a student would observe as they travel from the temperate rainforest to the woodlands.

After You Read the Module

Review Key Terms
Match the key terms on the left with the definitions on the right.

_____	1. Producers (Autotrophs)	a.	The energy captured by producers in an ecosystem minus the energy producers respire.
_____	2. Cellular respiration	b.	The process by which cells convert glucose into energy in the absence of oxygen.
_____	3. Anaerobic respiration	c.	The total amount of solar energy that producers in an ecosystem capture via photosynthesis over a given amount of time.
_____	4. Primary productivity	d.	Plants, algae, and some bacteria that use the Sun's energy to produce usable forms of energy, such as sugars.
_____	5. Gross primary productivity (GPP)	e.	The total mass of all living matter in a specific area.
_____	6. Net primary productivity (NPP)	f.	The amount of biomass present in an ecosystem at a particular time.
_____	7. Biomass	g.	The process by which cells unlock the energy of chemical compounds.
_____	8. Standing crop	h.	The rate of converting solar energy into organic compounds over a period of time.

MODULE 7: Trophic Levels, Energy Flow and the 10% Rule, Food Chains, and Food Webs

Before You Read the Module

Focus on Learning Goals
Use the module learning goals to guide your reading. On a separate piece of paper, write down each goal and take notes to help you meet each learning goal. After studying this module, you should be able to:
- 7-1 describe how energy and matter move through trophic levels in an ecosystem.
- 7-2 explain how low ecological efficiency causes energy to decrease at higher trophic levels.
- 7-3 explain why food webs experience feedback loops.

Key Terms

Consumer (Heterotroph)	Tertiary consumer	Decomposers
Herbivore (Primary consumer)	Trophic levels	Ecological efficiency
	Food chain	The 10% rule
Carnivore	Scavenger	Trophic pyramid
Secondary consumer	Detritivore	Food web

While You Read the Module
Answer the following questions as you read. Use a separate sheet of paper if necessary.

Module 7: Trophic Levels, Energy Flow and the 10% Rule, Food Chains, and Food Webs

Ecosystems depend on energy flowing through and matter cycling around trophic levels

1. Define consumers, or heterotrophs.

2. Define herbivores, or primary consumers, and give an example.

3. Define carnivores.

4. Define secondary consumers and give an example.

5. Define tertiary consumers and give an example.

6. Figure 7.1: Describe how the food chains are linked.

7. Define trophic levels.

8. Define food chain.

9. Describe why not all organisms can fit into a single trophic level.

10. Define scavengers and give an example.

11. Define detritivores and give an example.

12. Define decomposers.

13. Describe how scavengers, detritivores, and decomposers are important to an ecosystem.

14. Figure 7.2: Describe how visual representations can be used to show cycling and conservation of matter.

15. Explain why matter is cycled through an ecosystem, while energy flows through an ecosystem.

The flow of energy from one trophic level to the next has a low efficiency

Ecological Efficiency and Trophic Pyramids

16. Explain two reasons why not all the energy gets transferred to a higher trophic level.

17. Define ecological efficiency.

18. How much is the average ecological efficiency? Describe the 10% rule.

19. Define trophic pyramid.

20. Figure 7.4: Explain how the numbers change in the trophic pyramid.

21. Figure 7.4: Explain why the lion does not receive all 1,000 joules of energy from the zebra.

Ecological Efficiency Applied to Humans

22. What does the principle of ecological efficiency imply about the human diet?

23. Explain why more land is needed for humans to consume beef than is needed for humans to consume soybeans.

Food webs contain multiple, interconnected food chains with feedback loops

24. Define food web.

25. Figure 7.5: Why is this model more realistic than the one shown in Figure 7.1?

26. Figure 7.5: Explain the direction of arrows in the diagram.

27. Figure 7.5: Identify two omnivores and describe their different trophic levels within the ecosystem.

28. Figure 7.6: Describe the producers and consumers in an aquatic food web.

Feedback Loops

29. Describe why changes to the abundance of one species has the potential to impact another.

30. Visual Representation 1: The Yellowstone Ecosystem: Describe how the reintroduction of the wolves has impacted the population of cottonwood trees.

31. Visual Representation 1: The Yellowstone Ecosystem: Explain the flow of energy through the trophic levels found in the Yellowstone ecosystem.

Pursuing Environmental Solutions: The Practice of Precision Agriculture

32. Explain how loss of fertilizer affects farmers.

33. Describe precision agriculture.

34. Identify two advanced technologies that helped precision agriculture and describe how they have assisted in crop growth.

35. How can farmers justify or offset the large financial investment of precision agriculture?

Science Applied 1: Concept Explanation: Reversing Human Impacts on a Salty Lake

36. Where is Mono Lake?

37. What is a tufa tower?

38. Define terminal lake.

39. Describe the processes of salt formation that occur in Mono Lake.

40. What organisms can and can't tolerate the lake waters?

41. Describe the interactions of nutrient consumption that occur in the lake.

42. Why do the birds depend on the lake?

What happens when humans altered the natural water cycle?

43. Explain what the City of Los Angeles decided to do in 1941 with regard to Mono Lake.

44. Describe how the higher salinity that followed additional water withdrawals affected the Mono Lake ecosystem.

How can we reduce human impacts on the natural water cycle?

45. Identify the groups that caused a change at Mono Lake in 1994. Describe the change.

46. How did the city of Los Angeles decrease its water usage?

Where is the missing salt of Mono Lake?

47. Explain how Mono Lake never dried up or overflowed its banks.

48. Explain how Mono Lake contains less dissolved salt than predicted.

49. Why is the research on Mono Lake an excellent example of how environmental scientists make observations and develop hypotheses on how the natural world works?

After You Read the Module

Review Key Terms

Match the key terms on the left with the definitions on the right.

_____	1. Consumer (Heterotroph)	a. A carnivore that eats secondary consumers.
_____	2. Herbivore (Primary consumer)	b. Fungi and bacteria that complete the breakdown process by converting organic matter into small elements and molecules that can be recycled back into the ecosystem.
_____	3. Carnivore	c. Of the total biomass available at a given trophic level, only about 10 percent can be converted into energy at the next higher trophic level.
_____	4. Secondary consumer	d. A representation of the distribution of biomass, numbers, or energy among trophic levels.
_____	5. Tertiary consumer	e. A consumer that eats other consumers.
_____	6. Trophic levels	f. A consumer that eats producers.
_____	7. Food chain	g. The successive levels of organisms consuming one another.
_____	8. Scavenger	h. An organism that is incapable of photosynthesis and must therefore obtain its energy by consuming other organisms.
_____	9. Detritivore	i. A representation of the distribution of biomass, numbers, or energy among trophic levels.
_____	10. Decomposers	j. A carnivore that eats primary consumers.
_____	11. Ecological efficiency	k. The sequence of consumption from producers through tertiary consumers.
_____	12. The 10% rule	l. An organism that specializes in breaking down dead tissues and waste products into smaller particles.
_____	13. Trophic pyramid	m. The proportion of consumed energy that can be passed from one trophic level to another.
_____	14. Food web	n. An organism that consumes dead animals.

UNIT 1 Review Exercises

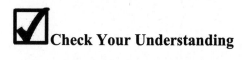Check Your Understanding

Review "Learning Goals Revisited" at the end of each module in Unit 1 of your textbook. Compare the notes you took while reading each module. Complete these exercises to review the unit. Use a separate sheet of paper if necessary.

1. Use the chart to compare mutualism, commensalism, and parasitism.

Relationship	Definition	Example
Mutualism		
Commensalism		
Parasitism		

2. Describe resource partitioning and give an example.

3. List the major terrestrial biomes on Earth. Include a brief description of the general climate and plant growth for each biome.

4. What are the major aquatic biomes on Earth?

5. Provide the formulas for photosynthesis and respiration. Describe how these processes lead to a steady state of the carbon cycle.

6. Briefly describe the steps of the nitrogen cycle.

7. Explain two ways in which excess phosphorus in the environment can result in environmental problems.

8. Examine Figure 7.4 on page 76, which shows the amount of energy that is present at each trophic level in the Serengeti ecosystem. What does the amount of energy at each level tell us about ecological efficiency?

9. Using an example, explain how changes in abundance of one species may potentially impact another species due to their interconnectedness.

Practice for Free-Response Questions

Complete these exercises to build and practice the skills you will need to answer free-response questions on the exam. Use a separate sheet of paper if necessary.

1. Bacteria may serve a variety of ecological functions.

    (a) An AP® Environmental Science student observes that in a particular rodent, a bacterial infection made the rat very ill and spread the illness to other rats in the same environment. Describe the type of relationship between the bacteria and the rat. Provide justification for your description.

    (b) A second AP® Environmental Science student observes that *Rhizobium* bacteria living on the roots of legumes resulted in plants that had increased crop production and overall health of the legume plant. Describe the type of relationship between the bacteria and the legume. Provide justification for your description.

2. Explain why terrestrial biomes found at the equator typically have higher biodiversity than those found at the poles.

3. Compare and contrast both the biotic and abiotic factors of the littoral and limnetic zone of a lake.

4. Explain the difference between the fast part of the carbon cycle and the slow part of the carbon cycle.

5. Explain how phosphorus results in dead zones.

6. Using the Visual Representation provided on pages 80-81, explain TWO environmental impacts of the reintroduction of wolves to Yellowstone.

Unit 1 Multiple-Choice Review Exam

Use the graph below to answer questions 1-3.

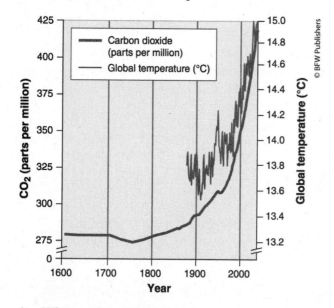

1. When carbon dioxide levels were at 325 ppm, what was the approximate global temperature in degrees Celsius?
   (a) 13.2
   (b) 14.1
   (c) 13.5
   (d) 13.8

2. If carbon dioxide levels hit 400 ppm, what do you estimate global temperature to be?
   (a) 14.8° Celsius
   (b) 13.8° Celsius
   (c) 17.2° Celsius
   (d) 2,100° Celsius

3. If the trend in global surface temperatures continues, what year do you estimate temperatures to become 15.0° Celsius?
   (a) 2000
   (b) 2100
   (c) 2500
   (d) 3000

4. Greenhouse gases in the atmosphere help
   (a) keep UV light from reaching Earth.
   (b) regulate temperatures near Earth's surface.
   (c) heat to be released back to space.
   (d) keep the ozone layer intact.

5. Which of the following accurately describes anthropogenic impacts to the carbon cycle?
   (a) Humans combust fossil fuels, releasing previously stored CO_2 into the atmosphere.
   (b) Deforestation results in increase in carbon sequestration
   (c) Use of CFCs allows for increased concentrations of carbon to be stored in sedimentary rock.
   (d) Impervious surfaces can be made with captured carbon to alter the water cycle.

6. If population of the United States is 307,000,000 and approximately 50 billion bottles of water are consumed in the United States each year, how many bottles are consumed per capita?
   (a) 7
   (b) 50
   (c) 100
   (d) 160

7. If there are 364,000 infants born every day and 152,000 people die each day, the net result is 212,000 new inhabitants on Earth each day. What is the approximate net gain of new inhabitants on Earth in one year?
   (a) 7 million
   (b) 77 million
   (c) 9.6 billion
   (d) 10.5 billion

8. Which is the correct equation for photosynthesis?
   (a) energy + $6H_2O$ + $7 CO_2$ → $C_6H_{12}O_6$ + $8O_2$
   (b) energy + $6 H_2O$ + $6 CO_2$ → $C_6H_{12}O_6$ + $6 O_2$
   (c) solar energy + $8 H_2O$ + $8 CO_2$ → $C_6H_{12}O_6$ + $12 O_2$
   (d) solar energy + $6 H_2O$ + $6 CO_2$ → $C_6H_{12}O_6$ + $6 O_2$

9. Which is the correct flow of energy in an ecosystem?
   (a) Producer → herbivore → carnivore → scavenger
   (b) producer → scavenger → herbivore → carnivore
   (c) scavenger → carnivore → herbivore → producer
   (d) scavenger → producer → Herbivore → carnivore

10. The net primary productivity of an ecosystem is 25 kg C/m^2/year, and the energy needed by the producers for their own respiration is 30 kg C/m^2/year. The gross primary productivity of such an ecosystem would be
    (a) 5 kg C/m^2/year.
    (b) 10 kg C/m^2/year.
    (c) 25 kg C/m^2/year.
    (d) 55 kg C/m^2/year.

11. An ecosystem has an ecological efficiency of 10 percent. If the tertiary consumer level has 1 kcal of energy, how much energy did the producer level contain?
    (a) 100 kcal
    (b) 900 kcal
    (c) 1,000 kcal
    (d) 10,000 kcal

12. Which of the following represents alterations humans make to the water cycle?
    (a) poor farming techniques have resulted in changing precipitation patterns
    (b) impervious surfaces alter aquifer recharge and surface runoff
    (c) use of rain gardens increases precipitation rates in certain areas
    (d) deforestation allows for greater evapotranspiration rates

13. _____Nitrogen fixation   (a) Nitrogen is assimilated into plant tissues.

14. _____Nitrification   (b) Organic matter is converted back into inorganic compounds.

15. _____Assimilation   (c) Ammonia is converted into nitrite and then nitrate.

16. _____Ammonification   (d) Bacteria convert ammonia into ammonium.

17. _____Mineralization   (e) Decomposers use waste as a food source and excrete
                                       ammonium.

18. Climate is
    (a) the amount of annual precipitation, the average temperature, and humidity.
    (b) the albedo and the average precipitation.
    (c) the conditions created by adiabatic cooling and heating.
    (d) the average weather that occurs in a given region over a long period of time.

19. Which of the following factors most directly results in a specific terrestrial biome?
    (a) particular flora and bacteria present
    (b) temperature and fauna present
    (c) temperature, precipitation, and altitude
    (d) precipitation and flora present

20. This terrestrial biome is characterized by cold winters and short growing seasons. The dominant
    plant life includes cone-bearing trees. Temperature in this region limits plant growth. Which
    biome fits this description?
    (a) the tundra
    (b) the boreal forest
    (c) the temperate seasonal forest
    (d) the cold desert

21. Which two terrestrial biomes are often used for agriculture because of high soil fertility?
    (a) tropical rainforests and temperate grasslands
    (b) tropical rainforests and temperate rainforests
    (c) temperate grasslands and temperate seasonal forests
    (d) boreal forest and temperate seasonal forests

22. Which describes ecosystem services associated with freshwater wetlands?
    (a) absorbing excess rainfall and filtering pollutants
    (b) preventing breeding for migrating birds and limiting fish nesting sites
    (c) emittance of carbon dioxide into the atmosphere
    (d) alteration of precipitation patterns to reduce flooding

23. Which describes predation?
    (a) One organism lives in or on another organism.
    (b) One animal kills and consumes another animal.
    (c) Two species have an interaction that benefits their chances of survival.
    (d) One species benefits another and the second species is neither harmed nor helped.

24. Which is an example of resource partitioning?
    (a) Wolves and foxes prey on rabbits.
    (b) Blue jays prevent smaller birds from eating at a bird feeder.
    (c) An orchid lives on a tree.
    (d) Wolves hunt at dawn and dusk and coyotes hunt at night.

25. Which of the following best describes why the tropical rainforest and seasonal forests have significantly higher NPP's than other terrestrial ecosystems?
    (a) High amounts of nutrients are leached from the soil due to the constant rains at the equator.
    (b) High winds allow for seed dispersal which allows the forests to spread, increasing overall NPP.
    (c) Larger populations of fauna result in high amounts of herbivory resulting in declining flora populations.
    (d) Higher temperatures and precipitation drive decomposition, adding nutrients for growing flora.

UNIT 2

The Living World: Biodiversity

Unit Summary

Unit 2 explores biodiversity and the importance of genetic, species, and habitat diversity. Higher biodiversity explains how ecosystems are resistant to disturbances, while providing services to the habitats, animals, plants, and humans. Learning and analyzing the importance of biodiversity allows scientists to recognize ecosystems that are healthy or in distress. The islands help us understand terrestrial type "islands" and determine possible stressors on the ecosystem. Recognizing change as a constant and natural process in the biodiversity of ecosystems can help scientists analyze the health of the system. Unit 2 lays the foundation for understanding the role of human activity in the loss of biodiversity and in determining realistic solutions. We must be able to identify and understand the changes in our world's ecosystems and how the negative changes alter the ecosystems and biodiversity.

MODULES IN THIS UNIT

Module 8: Introduction to Biodiversity
Module 9: Ecosystem Services
Module 10: Island Biogeography
Module 11: Ecological Tolerance
Module 12: Natural Disruptions to Ecosystems
Module 13: Adaptations
Module 14: Ecological Succession

Unit Opening Case: *The Benefits of Biodiversity*
Humans have long enjoyed the benefits provided by biodiversity including food, fiber for clothing, lumber, and medicines. Through data compiled from 67 studies that measured the number of species and ecosystem productivity from 600,000 sample locations around the world, researchers have shown that sites with more species have higher productivity than sites with fewer species. They concluded that ecosystems function more effectively with more biodiversity. It is critical that humans protect Earth's biodiversity so that we can continue to enjoy the many benefits it provides.

Do the Math
This unit contains the following "Do the Math" boxes to help prepare you for calculation questions you might encounter on the exam.
- "Converting Between Hectares and Acres" (page 100)
- "Estimating the Impact of the Founder Effect" (page 144)

To make sure you understand the concepts and techniques presented in this box, do the practice problems presented in the text as well as the additional "Practice the Math" problems that appear in Modules 8 and 13 of this study guide.

MODULE 8: Introduction to Biodiversity

Before You Read the Module

Focus on Learning Goals

Use the module learning goals to guide your reading. On a separate piece of paper, write down each goal and take notes to help you meet each learning goal. After studying this module, you should be able to:

- 8-1 identify how biodiversity exists at different scales.
- 8-2 explain how biodiversity affects responses to environmental stressors.
- 8-3 describe how we can calculate biodiversity.
- 8-4 describe how we estimate the number of species living on Earth.

Key Terms

Biodiversity   Habitat diversity   Species richness
Genetic diversity   Specialists   Species evenness
Population bottleneck   Generalists
Species diversity   Ecosystem diversity

While You Read the Module

Answer the following questions as you read. Use a separate sheet of paper if necessary.

Unit Opening Case: The Benefits of Biodiversity

1. Define biodiversity.

2. In what ways do humans benefit from species in nature?

3. What have scientists learned about the relationship between productivity in ecosystems and biodiversity?

Module 8: Introduction to Biodiversity

Biodiversity exists from genes to ecosystems

4. Identify the four levels of biodiversity.

5. Figure 8.1: Identify the how biodiversity differs among panels (a), (b), (c), and (d).

Genetic Diversity

6. Define genetic diversity.

7. Define population bottleneck.

Species Diversity

8. Describe a species.

9. Define species diversity.

Habitat Diversity

10. Define habitat diversity.

11. Define specialists and give an example.

12. Define generalists and give an example.

13. List the types of habitats in an area with a mixture of farms. Describe how this impacts diversity.

Ecosystem Diversity

14. Define ecosystem diversity.

15. Identify all the ecosystems in Los Angeles that allow for a high diversity of birds.

Increased biodiversity improves how populations and ecosystems respond to environmental stressors

Consequences of Genetic Diversity

16. What is an advantage of high genetic diversity?

17. Explain the consequence of low genetic diversity using cheetahs as the example.

Consequences of Habitat Diversity for Specialist and Generalist Species

18. Describe the different results for generalist and specialist species with the loss of habitat.

19. Describe the effect of continued habitat loss.

Consequences of Species Diversity

20. How is higher species diversity more productive and resilient?

21. Figure 8.4: Justify how soil fungi increases biomass of plant roots and shoots.

22. Describe how frog species may be an indicator of regional environmental health.

We can quantify biodiversity in terms of species richness and evenness

23. Define species richness.

24. Define species evenness.

25. Figure 8:6 Describe the differences between community 1 and 2.

26. Figure 8.6: Which community is considered more diverse and why?

It is difficult to estimate the number of species on Earth

27. To date, approximately how many species have been named?

28. Identify how many new species are discovered each year.

29. How do scientists use the number of insect species as an indicator of total number of species?

30. What is the current estimate for the number of species on Earth?

Practice the Math: Converting Between Hectares and Acres

Read "Do the Math: Converting Between Hectares and Acres" on page 100. Try the "Your Turn."
For more math practice, do the following problem. Remember to show your work. Use a separate sheet of paper if necessary.

2.5 acres = 1 hectare (ha)
1 acre = 0.40 ha

Convert the following from acres to hectares.
50,000 acres = _____ hectares
75,000 acres = _____ hectares
150,000 acres = _____ hectares

After You Read the Module

Review Key Terms
Match the key terms on the left with the definitions on the right.

_____	1. Biodiversity	a. Species that can live under a wide range of biotic or abiotic conditions.
_____	2. Genetic diversity	b. Species that only live under a narrow range of biotic or abiotic conditions.
_____	3. Population bottleneck	c. A measure of the genetic variation among individuals in a population.
_____	4. Species diversity	d. The relative proportion of individuals within the different species in a given area.
_____	5. Habitat diversity	e. The variety of habitats that exist in a given ecosystem.
_____	6. Specialists	f. The diversity of life forms in an environment.
_____	7. Generalists	g. When a large population declines in number, the amount of genetic diversity carried by the surviving individuals is greatly reduced.
_____	8. Ecosystem diversity	h. The variety of ecosystems that exist in a given region.
_____	9. Species richness	i. The number of species in a region or in a particular ecosystem.
_____	10. Species evenness	j. The number of different species in a given area.

MODULE 9: Ecosystem Services

Before You Read the Module

Focus on Learning Goals

Use the module learning goals to guide your reading. On a separate piece of paper, write down each goal and take notes to help you meet each learning goal. After studying this module, you should be able to:

- 9-1 explain the four categories of ecosystem services.
- 9-2 describe how human activities can disrupt ecosystem services.

Key Terms

Ecosystem services   Provision   Aquaculture

While You Read the Module

Answer the following questions as you read. Use a separate sheet of paper if necessary.

Module 9: Ecosystem Services

1. Define ecosystem services and give two examples.

2. Describe how an ecosystem service may directly and indirectly benefit people.

Ecosystem services include provisions, regulating services, support systems, and cultural services

3. Describe what it means for an ecosystem to have an intrinsic value.

4. List the four categories of ecosystem services.

Provisions

5. Define provision. Give an example.

6. Explain why a Pacific yew tree is such an important provision.

Regulating Services

7. Describe regulating services.

8. List two examples of a regulating service.

Support Systems

9. Describe the pollination support system.

Cultural Services

10. Describe cultural services.

11. Figure 9.4: Explain the cultural service shown in the photo.

12. Figure 9.4: Identify two possible activities that people could appreciate as a cultural service in the Grand Teton National Park.

The Monetary Value of Ecosystem Services

13. Explain the monetary replacement value of ecosystem services to the human economy.

Human activities can disrupt ecosystem services with economic and ecological consequences

14. How has the human population affected the trends of ecosystem services in the last 50 years?

Food Production

15. Describe how food production has impacted the loss of biodiversity.

16. Describe how food production has impacted the regulating service of carbon.

Fish and Shellfish Production

17. Define aquaculture.

18. Identify the effects of technological advances on the wild-caught fish and shellfish populations.

Water Availability

19. Explain what is happening to natural water sources in different parts of the world.

20. Figure 9.7: Describe the loss of water in Lake Mead.

21. In addition to having available water, identify what other water quality is needed for the resource to be drinkable for humans.

Pollination Services

22. Identify the causes for the decline of the bee population.

After You Read the Module

Review Key Terms
Match the key terms on the left with the definitions on the right.

_____	1. Ecosystem services	a. The farming of fish, shellfish, and seaweed.
_____	2. Provision	b. A good produced by an ecosystem that humans can use directly.
_____	3. Aquaculture	c. The processes by which life-supporting resources such as clean water, timber, fisheries, and agricultural crops are produced.

MODULE 10: Island Biogeography

Focus on Learning Goals

Use the module learning goals to guide your reading. On a separate piece of paper, write down each goal and take notes to help you meet each learning goal. After studying this module, you should be able to:

- 10-1 describe how island biogeography affects which species live on islands.
- 10-2 identify what determines the number of species on islands.
- 10-3 describe why some species on islands are particularly vulnerable to invasive species.

Key Terms

Island biogeography   Species-area curve

While You Read the Module

Answer the following questions as you read. Use a separate sheet of paper if necessary.

Module 10: Island Biogeography

Island biogeography affects the number of species living on islands and their ecological relationships

1. Define the theory of island biogeography.

The Species-area Curve for Islands

2. Define species-area curve.

3. Figure 10.2: Describe the relationship researchers found between island area and number of species.

4. What are the three reasons larger habitats contain more species?

Species-area Curves for Isolated Habitats in a Terrestrial Landscape

5. Figure 10.3: List the types of organisms that have similar species-area curve relationships in wetlands.

Effects of Island Size on Ecological Relationships

6. Explain why small islands can't support large predators.

The number of species on islands depends on rates of colonization and extinction

7. Explain how distance from habitat to colonizing species determines the species richness on an island.

Observing the Natural Effect of Island Distance

8. Figure 10.5: Describe the relationship of bird species numbers with large area islands compared to the near or far islands.

9. Figure 10.6: What is the effect of species richness when comparing mountain tops with larger areas and mountain tops that are closer to mountain ranges?

Experimentally Demonstrating the Effect of Island Distance

10. Describe the experiment of researchers E.O. Wilson and Daniel Simberloff and how this demonstrated fewer species on islands farther from the mainland.

A Model of Island Biogeography

11. Describe the model used to predict how many species an island will have.

12. Figure 10.9: Explain how the number of species change from graph (a) to (b).

13. Figure 10.10: Describe the locations where the highest and lowest number of species are found on islands.

Implications for Species Conservation

14. Describe the real-world application of understanding island biogeography and conserving biodiversity.

Some species on islands have evolved to be specialists and are vulnerable to invasive species

15. Explain how native species on islands have become vulnerable.

16. Explain how the brown tree snake of Guam impacted the native island animals.

17. Why is there a need to protect the biodiversity of birds on islands?

After You Read the Module

Review Key Terms
Match the key terms on the left with the definitions on the right.

_____   1. Island biogeography   a. The study of how species are distributed and
                                                interacting on islands.

_____   2. Species-area curve   b. A description of how the number of species on an
                                                island increases with the area of the island.

MODULE 11: Ecological Tolerance

Before You Read the Module

Focus on Learning Goals
Use the module learning goals to guide your reading. On a separate piece of paper, write down each goal and take notes to help you meet each learning goal. After studying this module, you should be able to:
- 11-1 explain what determines where individuals and species can live.
- 11-2 describe how environmental change affects species.
- 11-3 explain how environmental change can cause species extinctions.

Key Terms

Ecological tolerance (Fundamental niche)	Realized niche Geographic range	Mass extinction

While You Read the Module
Answer the following questions as you read. Use a separate sheet of paper if necessary.

Module 11: Ecological Tolerance

Ecological tolerance determines where individuals and species can live

1. Define ecological tolerance (fundamental niche).

2. List the abiotic conditions for the range of tolerance.

3. Figure 11.1: Where on the diagram is the best range of ecological tolerance?

4. Describe the fundamental niche.

5. Define realized niche.

6. Define geographic range.

7. Figure 11.2: Explain the realized niche of the red-winged blackbird compared to the yellow-headed blackbird.

Environmental change can alter the distribution of species

8. Figure 11.3: Describe how the distribution of the tree species changed from 18,000 years ago to present day.

9. What evidence shows the tree species migration?

10. Explain what could happen to the loblolly pine if temperatures increase in the future.

Environmental change beyond a species' ecological tolerance can cause species extinctions

11. What percentage of species that have ever lived on Earth are extinct?

12. Explain why scientists predict one-quarter of species in the Australian woodland and shrubland biomes will go extinct.

13. Explain why the loblolly pine may not move north in the United States.

The Five Global Mass Extinctions

14. Define mass extinction.

15. Figure 11.6: List the approximate geologic time for the five global mass extinctions starting with the oldest.

16. What mass extinction event had the greatest loss of genera and what percent was lost?

17. Identify the most known extinction event.

18. Explain the cause and effect of the Cretaceous extinction event.

The Sixth Mass Extinction

19. Why do scientists believe we are in the sixth mass extinction?

20. List the possible extinction rates from two scientific organizations and what we know for sure.

21. Describe what scientists agree to be the main causes of the mass extinction.

22. Describe what possible outcome the current environmental changes will have on species.

After You Read the Module

Review Key Terms
Match the key terms on the left with the definitions on the right.

_____ 1. Ecological tolerance (Fundamental niche)

a. The range of abiotic and biotic conditions under which a species actually lives.

_____ 2. Realized niche

b. The suite of abiotic conditions under which a species can survive, grow, and reproduce.

_____ 3. Geographic range

c. A large number of species that went extinct over a relatively short period of time.

_____ 4. Mass extinction

d. Areas of the world in which a species lives.

MODULE 12: Natural Disruptions to Ecosystems

Before You Read the Module

Focus on Learning Goals
Use the module learning goals to guide your reading. On a separate piece of paper, write down each goal and take notes to help you meet each learning goal. After studying this module, you should be able to:
- 12-1 describe how natural disruptions vary over time and space.
- 12-2 identify how Earth's climate and sea level have changed.
- 12-3 explain how natural disruptions can affect habitats and animal migrations.

Key Terms

Periodic disruption	Resistance	Intermediate disturbance
Episodic disruption	Resilience	hypothesis
Random disruption		

While You Read the Module

Answer the following questions as you read. Use a separate sheet of paper if necessary.

Module 12: Natural Disruptions to Ecosystems

Natural disruptions operate on a range of scales

Environmental Variation Over Time

1. Define periodic disruptions and give an example.

2. Define episodic disruptions and give an example.

3. Define random disruptions and give an example.

4. Figure 12.2: Explain why an ecosystem will have a limited chance for a forest fire if there was a fire this year.

5. Describe the cycling disturbance of disease outbreaks between hosts and disease-causing pathogens.

Duration and Spatial Extent of Disruptions

6. Figure 12.3: Explain the relationship between duration of time and spatial extent.

Ecosystem Resistance and Resilience

7. Define resistance.

8. Explain ecosystem high resistance.

9. Define resilience.

10. Give an example of high resilience.

11. Give an example of low resilience.

Earth's climate and sea level have changed over geologic time

Earth's Changing Climates

12. Describe how scientists use foraminifera to gather information about temperature changes.

13. Describe the significance of air bubbles trapped in ice.

14. Identify and describe how the two types of oxygen can indicate ancient temperatures.

15. Figure 12.6: Identify the two factors that are compared through the last 400,000 years.

16. Figure 12.6: Describe when the rapid declines in temperature occurred.

17. Figure 12.6: Identify the relationship between temperature and CO_2 concentrations.

Earth's Changing Sea Levels

18. Identify how scientists can indirectly determine sea level changes.

19. Figure 12.7: Describe how sea level fluctuated in the last 140,000 years.

20. Describe where the greatest ecosystem disruptions would have occurred.

Disruptions can cause large habitat changes and animal migrations

Large, Rare Disruptions

21. Explain how fires in California can stimulate growth.

22. Describe the need for fires in forests with longleaf pine.

23. Identify how forest managers control competing species in longleaf pine forests.

Intermediate Levels of Disruption

24. Define intermediate disturbance hypothesis.

25. Figure 12.9a: Analyze the graphed line in relation to each of the axis labels. Make note of lows and highs of the graphed line.

26. Figure 12.9b: Explain how an intermediate density of snails affects the species richness of algae.

Seasonal Disruptions Favoring Migrations

27. Identify the cause of seasonal animal migrations.

28. Describe the reason behind the long-term migration of bird species.

29. Figure 12.10: Explain the circular migration pattern of the Serengeti National Park.

After You Read the Module

Review Key Terms
Match the key terms on the left with the definitions on the right.

_____	1. Periodic disruption	a.	In an ecosystem, a measure of how much a disruption can affect the flows of energy and matter.
_____	2. Episodic disruption	b.	The hypothesis that ecosystems experiencing intermediate levels of disturbance will favor a higher level of diversity of species than those with high or low disturbance levels.
_____	3. Random disruption	c.	Occurring somewhat regularly, such as cycles of high rain and low rain that occur every 5 to 10 years.
_____	4. Resistance	d.	Occurring regularly, such as the cycles of day and night or the daily and monthly cycles of the moon's effect on ocean tides.
_____	5. Resilience	e.	The rate at which an ecosystem returns to its original state after a disruption.
_____	6. Intermediate disturbance hypothesis	f.	Occurring with no regular pattern, such as volcanic eruptions or hurricanes.

MODULE 13: Adaptations

Before You Read the Module

Focus on Learning Goals
Use the module learning goals to guide your reading. On a separate piece of paper, write down each goal and take notes to help you meet each learning goal. After studying this module, you should be able to:
- 13-1 describe how populations can evolve adaptations in response to environmental changes.
- 13-2 explain how evolution can occur over short and long timescales.

Key Terms

Evolution	Evolution by natural selection	Allopatric speciation
Microevolution	Fitness	Sympatric speciation
Macroevolution	Adaptation	Genetically modified organisms
Evolution by artificial selection	Evolution by random process	(GMO)

While You Read the Module

Answer the following questions as you read. Use a separate sheet of paper if necessary.

Module 13: Adaptations

Populations can evolve by artificial selection, natural selection, or random processes

1. Define evolution.

2. Define microevolution and give an example.

3. Define macroevolution.

Genes and Genetic Variation

4. Identify the role DNA plays for a given gene.

5. Describe how genotype and phenotype are different.

6. Describe one example of an environmental factor that determines phenotype.

7. Describe the example of a mosquito's genetic mutation, explain how this benefits an organism.

8. Describe how recombination of chromosomes could help the human immune system.

Evolution by Artificial Selection

9. What is the relationship between domesticated dogs and wolves?

10. How have breeders selectively bred different dog breeds?

11. Define evolution by artificial selection.

12. Figure 13.2: List the variety of food crops that originated from wild mustard.

13. Describe two unintended results of artificial selection.

Evolution by Natural Selection

14. Define evolution by natural selection.

15. Identify the two scientists who synthesized the theory of evolution by natural selection and the period in which they worked.

16. List the key ideas of Darwin's theory of evolution by natural selection.

17. Figure 13.3: Describe the size differences from one generation to the next in each of the generations of the amphipod.

18. Figure 13.3: Identify what caused the size of offspring to change through the generations.

19. Define fitness and give an example.

20. Define adaptation.

21. Figure 13.4: List and explain the adaptations of the desert plant in each photo.

Evolution by Random Processes

22. Define evolution by random processes.

23. List the five random processes of evolution.

24. Identify how mutations can be a disadvantage or an advantage to a population.

25. Describe gene flow.

26. Explain the decline of the Florida panther and the role gene flow played.

27. Describe genetic drift.

28. Figure 13.6: Describe the change in the genetic composition of a small population of mating mice; three white-haired and two black-haired. What could be a possible outcome?

29. Figure 13.6: Explain why a less-common genotype in a small population is more likely to be lost in future generations compared to a larger population.

30. Describe the bottleneck effect.

31. List the reasons a population may experience a drastic reduction in numbers.

32. Figure 13.7: After the bottleneck effect what features remain in the cheetah population?

33. How can the bottleneck effect lead to a decline in the population?

34. Explain the founder effect.

35. Figure 13.8: Imagine that a few individuals of a particular bird species happen to be blown off their usual migration route and land on a hospitable oceanic island. Describe the resulting differences between the mainland birds and the island birds.

36. Describe the founder effect in the Amish community in Pennsylvania.

Evolution can occur over short and long time scales by natural and artificial selection

Allopatric Speciation

37. Define allopatric speciation.

38. Figure 13.9: List the steps that occur to create allopatric speciation as shown in the figure.

39. What did Darwin discover about allopatric speciation and finches on the Galápagos Islands?

40. Figure 13.10: In the diagram how many finches are shown and from where did they originate?

41. List the differences among the groups of finches.

Sympatric Speciation

42. Define sympatric speciation.

43. Figure 13.11: List the types of wheat and their sets of chromosomes.

44. Identify the relationship between polyploid organisms and diploid organisms.

The Pace of Evolution

45. What is the timeframe for the evolution of the cichlid fishes of Lake Tanganyika?

46. Figure 13.12: In what ways do the cichlid fishes of Lake Tanganyika differ?

47. What is the timeframe for rapid evolution of the pupfish of the Death Valley region of California and Nevada?

48. Describe in detail how intensive fishing caused the Atlantic cod to rapidly evolve.

49. Define genetically modified organism (GMO).

50. Describe *Bacillus thuringiensis* and its importance.

51. What crops have scientists used *Bacillus thuringiensis* in?

52. What is the take home message of evolution and adaptations?

Practice the Math: Estimating the Impact of the Founder Effect

Read "Do the Math: Estimating the Impact of the Founder Effect" on page 144. Try the "Your Turn."
For more math practice, do the following problems. Remember to show your work. Use a separate sheet
of paper if necessary.

(a) A flowering weed colonizes a newly formed island in the Pacific. This weed species can have
purple, pink, and white flowers. Of the 600 plants, 450 are purple, 100 are white, and 50 are pink.
What is the percentage of each phenotype in the population?

(b) If the same weeds colonize another island, but 60 percent are pink, 35 percent are white, and 5
percent are purple, how many plants of each phenotype would you expect to see if there were 760
individuals on the new island?

Review Key Terms

Match the key terms on the left with the definitions on the right.

_____ 1. Evolution

a. The processes that alter the genetic composition of a population over time, but the changes are not related to differences in fitness among individuals.

_____ 2. Microevolution

b. The evolution of one species into two species, without any geographic isolation.

_____ 3. Macroevolution

c. Evolution that gives rise to new species, genera, families, classes, or phyla.

_____ 4. Evolution by artificial selection

d. The process in which humans determine which individuals to breed, typically with a preconceived set of traits in mind.

_____ 5. Evolution by natural selection

e. An individual's ability to survive and reproduce.

_____ 6. Fitness

f. The process in which the environment determines which individuals survive and reproduce.

_____ 7. Adaptation

g. A change in the genetic composition of a population over time.

_____ 8. Evolution by random processes

h. Evolution at the population level.

_____ 9. Allopatric speciation

i. An organism produced by copying genes from a species with some desirable trait and inserting them into other species of plants, animals, or microbes.

_____ 10. Sympatric speciation

j. The process of speciation that occurs with geographic isolation.

_____ 11. Genetically modified organism (GMO)

k. A trait that improves an individual's fitness.

MODULE 14: Ecological Succession

Before You Read the Module

Focus on Learning Goals
Use the module learning goals to guide your reading. On a separate piece of paper, write down each goal and take notes to help you meet each learning goal. After studying this module, you should be able to:
- 14-1 describe how terrestrial ecosystems experience succession.
- 14-2 describe whether succession occurs in aquatic ecosystems.
- 14-3 identify how succession impacts species richness, biomass, and productivity.
- 14-4 explain the importance of keystone species and indicator species.

Key Terms

Ecological succession	Secondary succession	Indicator species
Primary succession	Climax community	Endemic species
Pioneer species	Keystone species	Biodiversity hotspots

While You Read the Module
Answer the following questions as you read. Use a separate sheet of paper if necessary.

Module 14: Ecological Succession

Terrestrial ecosystems can experience primary or secondary succession

1. Define ecological succession.

Primary Succession

2. Define primary succession.

3. Figure 14.1: List the progression from rock to plant communities that develop as time increases during primary succession in this New England community.

4. Define pioneer species and give an example.

5. How does soil develop during primary succession?

6. What types of forest communities eventually develop in the East, Midwest, and Southwest of the United States? Explain why.

Secondary Succession

    7.   Define secondary succession.

    8.   When does secondary succession occur?

    9.   Figure 14.2: List the plant communities that develop as time increases for secondary succession in this New England forest.

    10. Describe how secondary succession progresses.

    11. Define climax community and give an example.

Succession also occurs in aquatic ecosystems

    12. List the steps of succession at a rocky intertidal zone.

    13. Describe what stream event could cause succession.

    14. Explain how the rapid recolonization of a stream occurs.

    15. Explain how a new lake becomes a terrestrial habitat.

    16. Figure 14.4: How many years does it take for lakes and ponds to become terrestrial habitats?

Succession initially increases species richness, total biomass, and productivity

    17. Figure 14.5: Describe the trend of the species richness over time in the graphs.

18. Figure 14.6: Describe the succession trends of the abandoned farm fields through time for each graph.

Keystone species alter species composition while indicator species indicate ecosystem characteristics

Keystone Species

19. Define keystone species.

20. Why is a beaver a good example of a keystone species?

21. Figure 14.8: Describe what happens to the arch when the keystone piece is removed.

22. Figure 14.9: Describe what researchers found about the effects of the sea star as a keystone species.

23. Give an example of how humans have impacted keystone species on islands in the South Pacific.

Indicator Species

24. Define indicator species.

25. Figure 14.10: Explain how the mayfly and red-cockaded woodpecker are indicator species.

26. Describe *E. coli* and explain why it is considered an indicator.

Visual Representation 2: Biodiversity in the Galápagos Islands

27. Explain how Island Biogeography impacts the colonization of the Galápagos Islands with plants and animals.

28. Describe the genetic diversity of the Darwin Finches.

29. Explain the adaptations that evolved for the marine iguana and the flightless cormorant.

30. Describe the specific plants that aided in the succession of the island.

Pursuing Environmental Solutions: Saving Marine Biodiversity Through the Power of Partnerships

31. What is the TNC and what is their mission?

32. What is the focus of The Nature Conservancy?

33. What role do shellfish play in an ecosystem?

34. How have shellfish been threatened? Explain the environmental effects that have occurred.

35. Why is The Nature Conservancy purchasing harvesting and exploitation rights?

36. List the areas or states where The Nature Conservancy is trying to assist in protecting biodiversity.

37. Describe bycatch.

38. How has TNC assisted in lowering percent of bycatch?

Science Applied 2: Concept Application: How Should We Prioritize the Protection of Species Diversity?

39. Define endemic species.

40. Give an example of a location for endemic species.

41. Define biodiversity hotspot.

42. How many hotspots have been identified?

What makes a hotspot hot?

43. What does Conservation International consider the criteria to be to qualify as a hotspot?

What else can make a hotspot hot?

44. What other two factors should be considered to make a hotspot?

What are the costs and benefits of conserving biodiversity hotspots?

45. How did the U.S. Fish and Wildlife Service help protect the California tiger salamander?

What about biodiversity coldspots?

46. Give an example of a biodiversity coldspot and explain.

47. Explain the value of biodiversity coldspots.

How can we reach a resolution?

48. Explain the balance that needs to be considered as we try to protect Earth's biodiversity.

After You Read the Module

Review Key Terms

Match the key terms on the left with the definitions on the right.

_____ 1. Ecological succession

_____ 2. Primary succession

_____ 3. Pioneer species

_____ 4. Secondary succession

_____ 5. Climax community

_____ 6. Keystone species

_____ 7. Indicator species

_____ 8. Endemic species

_____ 9. Biodiversity hotspots

a. Ecological succession occurring on surfaces with bare rock and no soil.

b. The succession of plant life that occurs in areas that have been disturbed but have not lost their soil.

c. Species that live in a very small area of the world and nowhere else, often in isolated locations such as the Hawaiian Islands.

d. The predictable replacement of one group of species by another group of species over time.

e. In primary succession, species that can survive with little or no soil.

f. A species that demonstrates a particular characteristic of an ecosystem.

g. Historically described as the final stage of succession.

h. Isolated areas that are home to so many endemic species, that they contain a high proportion of all the species found on Earth.

i. A species that is not very abundant but has large effects on an ecological community.

UNIT 2 Review Exercises

✓ **Check Your Understanding**

Review "Learning Goals Revisited" at the end of each module in Unit 2 of your textbook. Compare the notes you took while reading each module. Complete these exercises to review the unit. Use a separate sheet of paper if necessary.

1. Describe in your own words the measures between species evenness and species richness to determine biodiversity.

2. List and describe the five random processes that can drive evolution.

3. What factors determine the pace of evolution?

4. What five factors do scientists believe may be causing the sixth mass extinction?

5. Identify and describe the four ecosystem services.

6. Using the theory of island biogeography, describe what determines the number of species on larger islands away from the mainland.

7. Using ecological tolerance, explain how a living organism can be affected by moderate to extreme temperatures.

8. Describe how ecosystem resistance and resilience maintains the flow of energy in an ecosystem.

9. A forest fire occurs leaving only burnt vegetation and soil. Identify and describe the succession that will occur in a terrestrial environment recovering to the level of young pine development.

10. Identify and describe the type of succession that will occur in a terrestrial environment that is recovering from a landslide leaving only bare rock exposed to the level shrubbery growth.

Practice for Free-Response Questions

Complete these exercises to build and practice the skills you will need to answer free-response questions on the exam. Use a separate sheet of paper if necessary.

1. Identify and describe two different scales on which biodiversity can be measured.

2. See Figure 14.6 page 155: By examining farms that were abandoned at different times in the past, from 3 to 150 years ago, scientists could observe the plants on each farm.

   (a) Describe the changes that occurred in richness.

   (b) Describe the changes that occurred in biomass.

   (c) Describe the changes that occurred in net primary productivity.

3. Trees provide a variety of ecosystem services. Describe a service for each of the following.

   (a) Provisions

   (b) Regulating Services

   (c) Support Systems

   (d) Cultural Services

4. Rapid evolution is occurring in commercially harvested fish such as the Atlantic cod. Intensive fishing over several decades has continually targeted the largest adults. Describe two reasons the overharvested fish has been affected by rapid evolution.

5. See Figure 11.5 page 125: Explain how a species could go extinct from environmental conditions that may go through extreme changes.

Unit 2 Multiple-Choice Review Exam

1. Which statement describes the best possible habitat for a koala that is a species specialist?
    (a) The koala can consume any plant species in a forest.
    (b) The koala can only consume a single plant species, such as eucalyptus trees.
    (c) The koala can consume any plant species in prairies.
    (d) The koala is not limited to the type of plants it can consume.

2. Identify the type of diversity present if an area has a mixture of farms, pastures for cattle, abandoned farms, and forests with large trees.
    (a) genetic diversity
    (b) species diversity
    (c) habitat diversity
    (d) ecosystem diversity

3. Identify the type of diversity present if scientists observe that the white-tailed deer in colder locations have larger bodies compared to the deer living in Florida.
    (a) genetic diversity
    (b) species diversity
    (c) habitat diversity
    (d) ecosystem diversity

The following table represents the number of individuals of different species that were counted in three forest communities.

Species	Community A	Community B	Community C
Deer	75	20	10
Rabbit	1	20	10
Squirrel	1	20	10
Mouse	1	20	10
Chipmunk	1		10
Raccoon	1		10
Porcupine			10
Elk			10

4. Which statement best interprets this data?
    (a) Community A has greater species evenness and richness than Community B.
    (b) Community A has greater species richness than Community B.
    (c) Community B has greater species evenness than Community C.
    (d) Community C has greater species evenness and richness than Community A.

5. A cultural ecosystem service of Yellowstone National Park could best be described by which statement?
    (a) water and air purification by trees and soil
    (b) habitats for predators and prey
    (c) scenic overlooks for hikers to enjoy
    (d) harvesting timber to produce homes for humans

6. Which is an example of a provisional service?
   (a) bees producing honey for human consumption
   (b) trees providing air purification
   (c) bees pollinating wildflowers
   (d) research on trees and the input of carbon dioxide

7. Which statement describes an ecosystem service associated with freshwater wetlands?
   (a) Wetlands absorb some rainfall and sometimes slow flooding in nearby areas.
   (b) Wetlands recharge surface water and provide breeding habitats for local bird populations.
   (c) Wetlands provide critical breeding habitats for invasive species.
   (d) Wetlands absorb excess rainfall reducing flooding, filter pollutants from water, and provide critical breeding habitats for migratory birds.

8. According to the theory of island biogeography,
   (a) when conservation areas are close to each other, more species will persist.
   (b) species on islands far from the mainland are at the least risk of extinction.
   (c) multiple small conservation areas will protect species better than one large area of the same size.
   (d) edge habitat is important to protect for increased diversity.

9. According to the theory of Island biogeography, which island should have the highest number of species?
   (a) small islands close to the mainland
   (b) small islands far away from the mainland
   (c) large islands close to the mainland
   (d) large islands far away from the mainland

10. Ecological tolerance is
   (a) the abiotic conditions under which a species can survive, grow, and reproduce.
   (b) the areas of the world in which a species lives.
   (c) the limits to the abiotic conditions a species can tolerate.
   (d) the range of abiotic and biotic conditions under which a species actually lives.

11. The range of abiotic conditions and biotic conditions is known as the
   (a) geographic range.
   (b) fundamental niche.
   (c) ecological tolerance.
   (d) realized niche.

12. There is widespread agreement among scientists that Earth is in the _____ mass extinction due to human causes.
   (a) second
   (b) fourth
   (c) sixth
   (d) seventh

13. Which is true about natural ecosystem disruptions?
   (a) They are only caused by human-made events.
   (b) Episodic disruptions occur regularly, such as the cycles of day and night.
   (c) Random disruptions have an irregular pattern, similar to hurricanes.
   (d) Disruptions result with an increase in ecosystem productivity.

14. A low-intensity fire might kill some plant species, but fire-adapted species may benefit resulting in an ecosystem
    (a) resistance.
    (b) resilience.
    (c) disruption.
    (d) disturbance.

15. The intermediate disturbance hypothesis states that intermediate levels of disturbances will
    (a) increase ecosystem nutrient cycling.
    (b) decrease primary productivity.
    (c) increase species diversity.
    (d) decrease biomass.

16. Ancient air bubbles in ancient ice allow scientists to determine
    (a) oxygen concentrations of our current atmosphere.
    (b) changing sea levels.
    (c) past ocean oxygen levels.
    (d) temperatures of the distant past based on oxygen molecules.

17. Which of the following is an example of artificial selection?
    (a) Ostriches have lost the ability to fly.
    (b) Darwin's finches have beaks adapted to eating different foods.
    (c) Thoroughbred racehorses have been bred for speed.
    (d) Cichlids have diversified into nearly 200 species in Lake Tanganyika.

18. Which process is random?
    (a) birds from the mainland blown off course to an island
    (b) drug resistant bacteria develops in a hospital due to antibacterial cleaners
    (c) choosing the breed of dogs to have offspring
    (d) resistant weeds developing from overuse of an herbicide

19. Which of the following statements about mutations is correct?
    (a) All mutations are harmful.
    (b) Mutations can lead to less genetic diversity.
    (c) If mutations increase over time, evolution can occur.
    (d) Smaller populations have a better chance for mutations.

20. The cheetah's spot sizes changed in the next generation because of the reduction in genetic diversity due to the drastic reduction in population size, otherwise known as a
    (a) bottleneck effect.
    (b) gene flow.
    (c) genetic drift.
    (d) founder effect.

21. Speciation that occurs as a result of geographic isolation is called
    (a) gene flow.
    (b) allopatric speciation.
    (c) sympatric speciation.
    (d) bottleneck effect.

22. The Bt-corn is an example of
    (a) macroevolution.
    (b) evolution through geographic isolation.
    (c) evolution through genetic modification.
    (d) evolution through natural selection.

23. During primary succession
    (a) plants colonize soil that remains in areas that have been disturbed.
    (b) grasses and wildflowers colonize new areas.
    (c) early-arriving plants colonize bare rock.
    (d) pioneer species dominate because of their ability to survive in little to no soil.

24. Identify the correct sequence of succession events for a freshwater lake or pond.
    (a) Sedimentary rocks fill in the basin causing the lake to become shallow, plants colonize the lake, possibly hundreds to thousands of years later the lake turns into a terrestrial habitat.
    (b) Algae and aquatic plants colonize the body of water, sedimentary rocks erode into the basin causing the lake to become shallow, hundreds to thousands of years later turning into a terrestrial habitat.
    (c) Eventually the lake fills with aquatic vegetation developing into a terrestrial habitat.
    (d) Sedimentary rocks fill in the basin causing the lake to become shallow, algae colonize the lake, aquatic plants develop and fill the lake, possibly hundreds to thousands of years later the lake turns into a terrestrial habitat.

25. Indicator species could help identify which of the following ecosystem characteristics?
    (a) high quality trees to use as lumber
    (b) pollution indicators based on amount of lichens found on trees
    (c) species that have a large positive effect on an ecosystem
    (d) a species that has developed into a different species due to geographic isolation

26. Keystone species tend to
    (a) provide an essential service.
    (b) be a major energy producer.
    (c) have small effects on the ecosystem.
    (d) be the most abundant.

27. How many hectares are in 4,400 acres? (2.5 acres in 1ha)
    (a) 1.760
    (b) 17.60
    (c) 176.0
    (d) 1,760

28. A fundamental niche is
    (a) the abiotic conditions under which a species can survive, grow, and reproduce.
    (b) the areas of the world in which a species lives.
    (c) the limits to the abiotic conditions a species can tolerate.
    (d) the range of abiotic and biotic conditions under which a species actually lives.

29. Due to a drought, a lake dries into two small, independent lakes. Over time natural selection favors a different set of traits for a single species of fish found in both lakes. After a flood, the lakes are reconnected and fish populations rejoin but do not breed. Which process has occurred?
    (a) genetic drift
    (b) the bottleneck effect
    (c) sympatric speciation
    (d) allopatric speciation

30. When a beaver creates new habitat in an area, it is functioning as
    (a) a parasite.
    (b) an indicator species.
    (c) a mutualist.
    (d) a keystone species.

UNIT 3

Populations

Unit Summary

Many environmental factors can affect a population. This unit will examine the relationships that populations have with one another and how changes in the habitat affect populations. Species can be classified as specialist or generalist, each of which have unique reproductive strategies and respond differently to changes in the environment. Population growth is limited by factors such as space and resource availability, and different populations may respond differently to limiting factors. The human population has grown over time due to advancements in technology, medicine, agriculture, and the changing role of women. Mathematical calculations can be used to make predictions about future human populations. The demographic transition model explains how advancements in technology impact the size of a specific human population.

MODULES IN THIS UNIT

Module 15: Generalist and Specialist Species, *K*- and *r*-selected Species, and Survivorship Curves
Module 16: Carrying Capacity, Population Growth, and Resource Availability
Module 17: Age Structure Diagrams and Total Fertility Rates
Module 18: Human Population Dynamics and the Demographic Transition

---

Unit Opening Case: *New England Forests Contain a Variety of Species*
The case study illustrates the complexity of community interactions and provides a historical example of ecological recovery and secondary succession. The story follows the New England ecosystem through a series of changes including deforestation and conversion to croplands, field abandonment, and the reestablishment of grasses and trees. This story exemplifies the multifaceted interactions among species that drive the ever-changing ecosystem composition.

Do the Math
This unit contains the following "Do the Math" boxes to help prepare you for calculation questions you might encounter on the exam.
- "Calculating Population Growth" (page 195)
- "Comparing Population Growth in Two Countries" (page 207)

To make sure you understand the concepts and techniques presented in these boxes, do the practice problems presented in the text as well as the additional "Practice the Math" problems that appear in Module 17 and Module 18 of this study guide.

MODULE 15: Generalist and Specialist Species, *K* - and *r* -selected Species, and Survivorship Curves

Focus on Learning Goals
Use the module learning goals to guide your reading. On a separate piece of paper, write down each goal and take notes to help you meet each learning goal. After studying this module, you should be able to:
- 15-1 identify the differences between generalist and specialist species.
- 15-2 describe the differences between *K*- and *r*-selected species.
- 15-3 explain survivorship curves.

Key Terms

Population growth rate (Intrinsic growth rate)	Carrying capacity	Survivorship curve
	r -selected species	Type I survivorship curve
Biotic potential	Overshoot	Type II survivorship curve
K -selected species	Dieback (Die-off)	Type III survivorship curve

While You Read the Module

Answer the following questions as you read. Use a separate sheet of paper if necessary.

Unit Opening Case: New England Forests Contain a Variety of Species

1. Why did Massachusetts experience deforestation from 1620 to the 1800s?

2. When people left their New England farms what happened to the farmland?

3. What became the dominant plant during this transformation? What did this plant attract?

4. How did the goldenrod and beetle interaction create a changing ecosystem?

5. Explain the changes that followed through the 1900s.

6. What does this case illustrate?

Module 15: Generalist and Specialist Species, *K*- and *r*-selected Species, and Survivorship Curves

Generalists exist under a broad range of conditions while specialists exist under a narrow range

7. What two factors affect the abundance of species in a particular location or habitat?

8. Describe the difference between niche generalists and niche specialists.

9. When do niche specialists tend to do well? When are they vulnerable?

10. When do niche generalists tend to fare better than specialists?

11. Which type of species is more resilient to global change?

Species have different reproductive strategies

12. Define population growth rate/ intrinsic growth rate.

13. Define biotic potential.

K-selected Species

14. Define *K*-selected species.

15. Define carrying capacity. How is it denoted?

16. What determines the population size of a *K*-selected species? Does it fluctuate?

17. Provide four characteristics of *K*-selected species.

18. What is an example of a *K*-selected species?

19. Why does the slow growth of *K*-selected species pose a challenge?

20. Why do *K*-selected species have stable populations?

21. Describe why *K*-selected species may be vulnerable to invasive species.

r-selected Species

22. Define *r*-selected species.

23. Provide two characteristics of *r*-selected species.

24. Define the process of overshoot and dieback/die-off.

25. Figure 15.2: What happens to a population that exceeds its carrying capacity?

26. Table 15.1: Seeing and writing the table will help with your understanding. Copy the table into your notes.

27. Create a memory device to differentiate between *K*-selected species and *r*-selected species.

28. Provide at least four characteristics for *r*-selected species.

29. What is an example of an *r*-selected species?

30. Why do the populations of *r*-selected species fluctuate so widely?

31. Do most species fall toward the end of the *r*-selected versus *K*-selected table? Provide an example.

Species exhibit three distinctly different survivorship curves

32. Define survivorship curves.

33. Define a type I survivorship curve.

34. Give an example of a species with a type I survivorship curve.

35. Define type II survivorship curve.

36. Give an example of a species with a type II survivorship curve.

37. Define type III survivorship curve.

38. Give an example of a species with a type III survivorship curve.

39. Figure 15.3: Describe the typical survivorship curves of a *K*-selected species and an r-selected species.

40. What three factors contribute to the survival and success of a given species?

Review Key Terms

Match the key terms on the left with the definitions on the right.

_____	1. Population growth rate (Intrinsic growth rate)	a. A species that has a high intrinsic growth rate, and their population typically increases rapidly.
_____	2. Biotic potential	b. A pattern of survival over time in which there is low survivorship (a high death rate) early in life with few individuals reaching adulthood.
_____	3. K-selected species	c. When a population becomes larger than the environment's carrying capacity.
_____	4. Carrying capacity	d. A pattern of survival over time in which there is high survival throughout most of the life span, but then individuals start to die in large numbers as they approach old age.
_____	5. r-selected species	e. The limit to the number of individuals that can be supported by an existing habitat or ecosystem, and is denoted as K.
_____	6. Overshoot	f. A species with a low intrinsic growth rate that causes the population to increase slowly until it reaches the carrying capacity of the environment.
_____	7. Dieback (Die-off)	g. A pattern of survival over time in which there is a relatively constant decline in survivorship throughout most of the life span.
_____	8. Survivorship curve	h. A graph that represents the distinct patterns of species survival as a function of age.
_____	9. Type I survivorship curve	i. A rapid decline in a population due to death.
_____	10. Type II survivorship curve	j. The number of offspring an individual can produce in a given time period, minus the deaths of the individual or its offspring during the same period.
_____	11. Type III survivorship curve	k. Under ideal conditions with unlimited resources available, every population has a maximum potential for growth.

MODULE 16: Carrying Capacity, Population Growth, and Resource Availability

Before You Read the Module

Focus on Learning Goals
Use the module learning goals to guide your reading. On a separate piece of paper, write down each goal and take notes to help you meet each learning goal. After studying this module, you should be able to:
- 16-1 explain the exponential growth model, and why it produces a J-shaped curve.
- 16-2 describe how the logistic growth model explains carrying capacity and produces an S-shaped curve.
- 16-3 justify how carrying capacity can have an impact on ecosystems.
- 16-4 identify how population can decline below carrying capacity.

Key Terms

Density-dependent factor	Fecundity	Logistic growth model
Density-independent factor	Exponential growth model	S-shaped curve
Population growth models	J-shaped curve	Limiting resource

While You Read the Module

Answer the following questions as you read. Use a separate sheet of paper if necessary.

Module 16: Carrying Capacity, Population Growth, and Resource Availability

1. What usually limits the growth of a population?

2. Define density-dependent factors and give an example.

3. Define density-independent factors.

The exponential growth model describes populations that continuously increase, and is the first step in explaining carrying capacity

4. Define population growth models.

5. Define fecundity.

6. Define exponential growth model.

7. What does the equation in the exponential growth model tell us?

8. Define J-shaped curve.

9. Describe how the J-shaped curve develops.

10. Figure 16.1: Draw a J-curve in your notes or on your paper. Describe the growth it represents and how that growth happens.

The logistic growth model describes populations that are limited by the carrying capacity

11. Define the logistic growth model.

12. Figure 16.2: Draw and label the logistic growth model.

13. Define an S-shaped curve.

14. Why are logistic growth models used to predict how populations will respond to density-dependent factors but not density-independent factors?

Density-Dependent Factors

15. Identify what Gause suspected was the limiting factor for growing *Paramecium*.

16. Figure 16.3: Compare and contrast the two graphs, (a) low-food supply and (b) high-food supply.

17. Define limiting resource.

18. List examples of limiting resources for terrestrial plant populations and animal populations.

19. Figure 16.3: Identify the graph with the largest carrying capacity (K) for Gause's experiment.

Density-independent Factors

20. List examples of density-independent factors.

Carrying capacity can impact ecosystem diebacks

21. List the sequence of events for the overshoot and die-off of the reindeer population on St. Paul Island in Alaska.

22. Describe what happens when populations are not managed properly. Is this reversible? Explain why or why not.

23. Figure 16.6: Explain what is happening in population oscillations.

24. Figure 16.7: Explain how the predator-prey relationship affects population growth using the example of snowshoe hare and lynx populations.

25. Figure 16.8: Describe where the overshoot and die-off occur with the wolves and moose. Include the reason each event occurred.

Populations change when a resource base shrinks

26. Provide an example in which an organism may be limited by space rather than food.

After You Read the Module

Review Key Terms
Match the key terms on the left with the definitions on the right.

_____   1. Density-dependent factor

a.   A resource that a population cannot live without and that occurs in quantities lower than the population would require to increase in size.

_____   2. Density-independent factor

b.   Mathematical equations that can be used to predict population size at any moment in time.

_____   3. Population growth models

c.   The ability to produce an abundance of offspring.

_____   4. Fecundity

d.   A factor that has the same effect on an individual's probability of survival and reproduction at any population size.

_____   5. Exponential growth model

e.   A growth model that describes a population whose growth is initially exponential, but slows as the population approaches the carrying capacity of the environment.

_____   6. J-shaped curve

f.   The curve of the exponential growth model when graphed.

_____   7. Logistic growth model

g.   A factor that influences an individual's probability of survival and reproduction in a manner that depends on the size of the population.

_____   8. S-shaped curve

h.   The shape of the logistic growth model when graphed.

_____   9. Limiting resource

i.   A growth model that estimates a population's future size after a period of time based on the biotic potential and the number of reproducing individuals currently in the population.

MODULE 17: Age Structure Diagrams and Total Fertility Rates

Before You Read the Module

Focus on Learning Goals

Use the module learning goals to guide your reading. On a separate piece of paper, write down each goal and take notes to help you meet each learning goal. After studying this module, you should be able to:

- 17-1 explain the factors that justify the carrying capacity of humans on Earth.
- 17-2 describe the social and cultural drivers of human population change.
- 17-3 describe and interpret age structure diagrams.
- 17-4 explain the factors that influence the total fertility rate.
- 17-5 describe replacement level fertility.

Key Terms

Demography	Life expectancy	Developed countries
Demographer	Infant mortality	Population momentum
Immigration	Child mortality	Total fertility rate (TFR)
Emigration	Environmental justice	Family planning
Crude birth rate (CBR)	Age structure diagram	Replacement-level fertility
Crude death rate (CDR)	Population pyramid	
Net migration rate	Developing countries	

While You Read the Module

Answer the following questions as you read. Use a separate sheet of paper if necessary.

Module 17: Age Structure Diagrams and Total Fertility Rates

The total number of people the Earth can support is not known

1. How much does the Earth's human population increase every 5 days?

2. Figure 17.1: Looking at the graph, when did human population start to dramatically change?

3. What change occurred 400 years ago that allowed the population to grow so much faster?

4. How did Thomas Malthus view Earth's carrying capacity?

5. What do other scientists believe about the Earth's carrying capacity?

6. Figure 17.2a: Describe the significance of the purple bars on the graph.

7. Explain what innovations led to a food surplus.

Many factors drive human population change

8. Define demography.

9. Define demographer.

Social and Cultural Factors in Population Change

10. Figure 17.3: Consider the human population as a system. Describe the inputs and outputs of that system.

11. Define immigration.

12. Define emigration.

13. Define crude birth rate (CBR).

14. Define crude death rate (CDR).

15. What was Earth's global population growth rate in 2020?

16. Write the equation for the global population growth rate in percent.

17. Define net migration rate.

18. Write the equation for the national population growth rate in percent.

19. Describe how both the United States and Canada will experience net population growth over the next few decades.

20. Describe how Georgia, a country in western Asia, will have a population decrease.

21. Define life expectancy and describe what it generally implies.

22. What are the three ways life expectancy is reported?

23. What was overall life expectancy in the United States in 2020? What was life expectancy for men and for women? How do demographers predict the Covid-19 pandemic will affect life expectancy in the United States?

24. Why is there a discrepancy between the life expectancy in men and women? Why will the gap of life expectancy decrease between men and women over time?

25. Define infant mortality.

26. Define child mortality.

27. Finish the following statement: "If a country's life expectancy is relatively high and its infant mortality rate is relatively low…"

28. What is the global infant mortality rate? List some countries and their known infant mortality rates.

29. Why does the United States show a variation in infant mortality rates?

30. Define environmental justice.

31. Why would a country with high life expectancy have a high crude death rate?

32. According to the World Health Organization what are the top two killers worldwide?

33. Which transmissible disease has the greatest effect on deaths worldwide?

34. How have death rates for HIV changed over time? Why does HIV have a more disruptive effect than other illnesses?

35. Figure 17.6: Worldwide, where are most people living with HIV?

36. How has Covid-19 impacted deaths worldwide?

Age structure diagrams describe how populations are distributed across age ranges and provide insights into future population changes

37. Define age structure diagrams.

38. Figure 17.7: What differentiates the left and right sides of the age structure diagrams?

39. Figure 17.7: What does each horizontal bar on the age structure diagrams indicate?

40. Figure 17.7: What is the label at the bottom of the age structure diagrams, the *x*-axis?

41. Figure 17.7: What does the total area of all the bars represent on the age structure diagrams?

42. Figure 17.7: Examine the population size of each diagram. Identify the countries with the largest and the smallest populations in the figure.

43. Define population pyramid.

44. Define developing countries.

45. Define developed countries.

46. Describe what a population pyramid indicates about the population of that country. Give an example of a country with a population pyramid.

47. If a country has little difference between the number of individuals in younger age groups and in older groups, how will its age structure diagram appear? Give an example.

48. What are characteristics of a population that has an inverted pyramid age structure diagram? What does that age structure diagram shape imply about a country? Provide an example.

49. Define population momentum.

50. Explain why population momentum has been compared to a long, heavy freight train.

51. Create a table that demonstrates the implications of each of the different types of age-structure diagrams.

Age when having the first child and access to family planning influence the total fertility rate

52. Define total fertility rate (TFR). What do demographers consider to be childbearing years?

53. What was the total fertility rate (TFR) for the United States in 2020?

54. Explain the factors influence TFR.

55. Figure 17.8: Make a claim about the impact that GDP per capita has on the number of children each woman has.

56. Define family planning.

57. Figure 17.9: Make a claim with evidence about the TFR of educated women and those lacking formal education. Provide reasoning for your claim.

Replacement level fertility is the TFR that leads to a stable population

58. Define replacement-level fertility.

59. Why is replacement-level fertility typically higher than two?

60. Why are there differences between replacement level fertility in developing and developed countries?

Practice the Math: Calculating Population

Read "Do the Math: Calculating Population Growth," on page 195. Try "Your Turn." For more math practice, do the following problem. Remember to show your work. Use a separate piece of paper if necessary.

In 2014, Lebanon had a population of 5.9 million people. The crude birth rate was 15, while the crude death rate was 5. The net migration rate was 84 per 1,000. How many people will be added to the population if this trend continues over the next year?

After You Read the Module

Review Key Terms

Match the key terms on the left with the definitions on the right.

_____ 1. Demography

_____ 2. Demographer

_____ 3. Immigration

_____ 4. Emigration

_____ 5. Crude birth rate (CBR)

_____ 6. Crude death rate (CDR)

_____ 7. Net migration rate

_____ 8. Life expectancy

_____ 9. Infant mortality

_____ 10. Child mortality

_____ 11. Environmental justice

_____ 12. Age structure diagram

_____ 13. Population pyramid

_____ 14. Developing countries

_____ 15. Developed countries

_____ 16. Population momentum

_____ 17. Total fertility rate (TFR)

_____ 18. Family planning

_____ 19. Replacement-level fertility

a. The average number of years that an infant born in a particular year in a particular country can be expected to live, given the country's average life span and death rate.

b. Countries with relatively low levels of industrialization and income.

c. The total fertility rate required to offset the average number of deaths in a population in order to maintain the current population size.

d. The number of deaths of children under age 5 per 1,000 live births.

e. The number of births per 1,000 individuals per year.

f. The number of deaths of children under 1 year of age per 1,000 live births.

g. An estimate of the average number of children that each woman in a population will bear throughout her childbearing years.

h. A visual representation of the number of individuals within specific age groups for a country, typically expressed for males and females.

i. Continued population growth after growth reduction measures have been implemented.

j. An age structure diagram that is widest at the bottom and smallest at the top, typical of developing countries.

k. The movement of people out of a country or region.

l. Countries that have relatively high levels of industrialization and income.

m. The study of the disproportionate exposure to environmental hazards experienced by people of color, recent immigrants and people of lower socio-economic backgrounds; and is both an academic field and a social movement.

n. A scientist in the field of demography.

o. The study of human populations and population trends.

p. The difference between immigration and emigration in a given year per 1,000 people in a country.

q. The regulation of the number or spacing of offspring through the use of birth control.

r. The number of deaths per 1,000 individuals per year.

s. The movement of people into a country or region, from another country or region.

MODULE 18: Human Population Dynamics and the Demographic Transition

Before You Read the Module

Focus on Learning Goals
Use the module learning goals to guide your reading. On a separate piece of paper, write down each goal and take notes to help you meet each learning goal. After studying this module, you should be able to:
- 18-1 explain what factors lead to population growth and decline.
- 18-2 calculate doubling times using the rule of 70.
- 18-3 describe the demographic transition.
- 18-4 explain how population and consumption interact to influence the environment.

Key Terms

Doubling time	Theory of demographic	IPAT equation (p. 211)
Rule of 70	transition	

While You Read the Module

Answer the following questions as you read. Use a separate sheet of paper if necessary.

Module 18: Human Population Dynamics and the Demographic Transition

Social, political, and economic factors contribute to population growth and decline

1. What is the approximate population of China?

2. What is the approximate population of the United States?

3. Describe how China and the United States have changed over time in regards to consumption of resources and production of pollutants.

Limiting Population Growth in China

4. Explain the history of the "one-child" policy in China.

5. What is China's TFR? Compare India's population to China.

6. Explain the implication for resource consumption of China's growing middle class. List factors that will be affected.

7. How has car usage in China changed and what are the environmental implications?

8. Explain why the United States may be concerned about China's air pollution.

Factors of Global Population Change

9. What was Earth's population in 2022 and its growth per day?

10. How does having a lower infant mortality for a number of years impact a family's decision to have fewer children?

11. What is the cause of the increase in density-independent factors influencing the human population?

12. List density-dependent factors that affect the size of the human population.

Largest Cities in the World

13. How has the number of people living in urban areas changed over time?

14. Table 18.2: Are the majority of large urban areas in developed or developing countries?

The rule of 70 allows you to calculate the doubling time

15. Define doubling time.

16. Write the equation for doubling time.

17. What is the doubling time for a population that is growing at two percent?

The demographic transition is a model that predicts population changes as countries develop

The Theory of Demographic Transition

18. Define theory of demographic transition.

19. Figure 18.3: List the names of the phases of the demographic transition. What is each phase also referred to as?

20. List the characteristics of phase 1: slow population growth.

21. Give an example of a phase 1 society.

22. List the characteristics of phase 2: rapid population growth.

23. Give an example of a phase 2 society.

24. List the characteristics of phase 3: stable population growth.

25. Give an example of a phase 3 society.

26. List the characteristics of phase 4: declining population growth.

27. Give four examples of a phase 4 society.

28. What do demographers think might happen after TFR reaches a low point of 1.2 or 1.5?

Population and consumption influence the environment differently, depending on the country

29. What are the two critical factors that determine the impact humans have on Earth?

Resource Use

30. How many humans on Earth live in developed and developing countries?

31. Figure 18.5: List the twelve countries by population size from largest to smallest. For each country, identify whether it is developed or developing.

The IPAT Equation

32. Figure 18.7: Contrast the material possessions featured in each of the two photographs.

33. Define the IPAT equation.

34. Describe the population factor of the impact (IPAT) equation.

35. Describe the affluence factor of the impact (IPAT) equation.

36. Describe the technology factor of the impact (IPAT) equation.

37. Describe destructive technology.

Practice the Math: Comparing Population Growth in Two Countries

Read "Do the Math: Comparing Population Growth in Two Countries" on page 207. Try "Your Turn." For more math practice, do the following problem. Remember to show your work. Use a separate sheet of paper if necessary.

In 2017 Angola had a growth rate of 3.52 percent and India had a growth rate of 1.17 percent.

(a) How long will it take for Angola and India to double in population?

(b) Which country has the faster doubling time? How many doubling periods will that country go through in the time it takes the slower country to double once?

Visual Representation 3 Population dynamics in Japan and Tanzania

38. Describe the age structure diagram of Japan and provide demographic statistics (TFR, life expectancy, and other factors) that support the shape you described.

39. Why does the Sika deer population seem to stabilize at 1200 individuals?

40. How does Tanzania's location within the demographic transition give rise to the shape of the population pyramid it exhibits? Support your statement with demographic evidence provided.

Pursuing Environmental Solutions

Gender Equity and Population Control in Kerala

41. What is the growth rate of India?

42. What method did India attempt to use as population control?

43. Why did the population of Kerala, India start growing faster than the rest of the country?

44. What is the fertility rate of Kerala compared to the rest of India?

45. What did Kerala emphasize for responsible population growth?

46. What is the evidence that Kerala's three E's helped to drop the fertility rate?

47. List other similarly impoverished countries that have reduced fertility rates.

48. What is an important component of the United Nations Program?

Science Applied 3: How Can We Control Overabundant Animal Populations?

What causes animal population explosions?

49. List the reasons for the population increase of the white-tailed deer.

50. What kind of feedback loop has been portrayed with the white-tailed deer population?

51. List other species in North America with drastically increasing populations.

What effects do overabundant species have on communities?

52. List the effects of overabundant species.

53. Why are automobile collisions with animals such a negative aspect? Think beyond the reading.

How can we control overabundant species?

54. How can we control overabundant species locally?

55. Describe how the culling of an animal herd could take place.

56. Who tends to oppose culling herds of animals?

57. What strategy to control overabundant species has more public support? What are some of the methods to distribute this strategy?

58. Identify some of the positive and negative aspects of animal birth control.

59. What should society consider before putting any processes into practice for the control of overabundant populations?

After You Read the Module

Review Key Terms
Match the key terms on the left with the definitions on the right.

_____ 1. Doubling time	a. A method which dictates that by dividing the number 70 by the percentage population growth rate we can determine a population's doubling time.
_____ 2. Rule of 70	b. A conceptual representation of the three major factors that influence environmental Impact: Population of humans, Affluence, Technology.
_____ 3. Theory of demographic transition	c. The number of years it takes a population to double.
_____ 4. IPAT equation	d. A theory that states that a country moves from high to lower birth and death rates as development occurs and that country moves from a preindustrial to an industrialized economic system.

UNIT 3 Review Exercises

Check Your Understanding

Review "Learning Goals Revisited" on pages 179, 188, 201, and 214 of your textbook. Compare the notes you took while reading each module. Complete these exercises to review the chapter. Use a separate piece of paper if necessary.

1. Distinguish between generalist and specialist, *r*- and *K*-selected species, and Type I, II, and III survivorship curves.

2. Draw both a J and S curve and label each. Explain what distinguishes an S-curve from a J curve.

3. Distinguish between density dependent factors and density independent factors.

4. Draw the three different types of age structure diagrams and explain what they imply about each population.

5. Using the growth rate formula, calculate the growth rate of a country with a CBR of 15, a CDR of 10, an immigration rate of 5, and an emigration rate of 2.

   Formula:

   $$\text{growth rate} = \frac{(\text{crude birth rate} + \text{immigration}) - (\text{crude death rate} + \text{emigration})}{10}$$

6. Calculate the doubling time of the country in question 5.

   Formula:

   $$\text{doubling time (in years)} = \frac{70}{\text{growth rate}}$$

7. Sketch the demographic transition model. Label the four stages and describe what is happening in each to influence birth rate, death rate, and total population.

Practice for Free-Response Questions

Complete this exercise to build and practice the skills you will need to answer free-response questions on the exam. Use a separate sheet of paper if necessary.

1. Examples of *K*-selected species include whales and elephants. Describe one characteristic of *K*-selected species and describe how that characteristic makes these species more prone to either extinction OR invasive species.

2. Using the data found in Figure 16.8 on page 186, make a claim with evidence about the wolf population in 1980 and explain WHY the moose population changed between 1989 and 2014.

3. Refer to Figure 17.5 on page 194 and Figure 17.8 on page 199.
    (a) Describe TWO economic reasons that infant mortality rate is lowest in Japan.

    (b) Several factors influence infant mortality rate.
        i.   Describe how education of women influences infant mortality rate.

        ii. Describe how access to health care and sanitation influences infant mortality rate.

4. Using Figure 17.7 on page 197, explain which stage of the demographic transition model Nigeria (a) might be in compared to Germany (c).

Unit 3 Multiple-Choice Review Exam

1. Which of the following characteristics makes generalist species more likely to thrive in a changing environment?
   (a) They have a specific food source that may be threatened by a changing environment.
   (b) They occupy a narrow niche and have specific habitat requirements that are threatened by a changing environment.
   (c) They have a wide range of tolerance and can tolerate a variety of environmental conditions.
   (d) They require specific abiotic components are not generally able to survive in alternative environmental conditions.

2. If a population of 300 deer increases to 400 deer, the percent change is
   (a) 33 percent.
   (b) 300 percent.
   (c) 50 percent.
   (d) 25 percent.

3. Which is an *r*-selected species?
   (a) humans
   (b) elephants
   (c) cockroaches
   (d) dogs

4. Which refers to the maximum growth potential of a population under ideal conditions with unlimited resources?
   (a) intrinsic growth rate
   (b) exponential growth
   (c) carrying capacity
   (d) logistic growth

Use the following figure for questions 5-7.

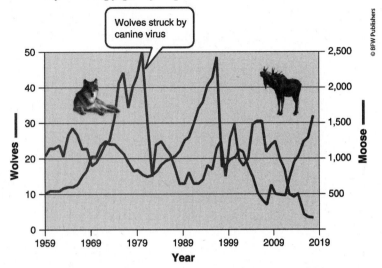

Data from R. O. Peterson and J. A. Vucetich, Ecological Studies of Wolves on Isle Royale: Annual Report 2016–2017, School of Forest Resources and Environmental Science, Michigan Technological University.

5. In what year did the wolf population dramatically decrease due to a canine virus?
   (a) 1969
   (b) 1980
   (c) 1989
   (d) 2011

6. The graph illustrates
   (a) a mutualistic relationship.
   (b) logistic growth.
   (c) population oscillations.
   (d) survivorship curves.

7. Which of the following is best supported by the graph of wolves and moose on Isle Royale from 1959 to 2019?
   (a) There is a direct relationship. As wolf populations increase, so do moose.
   (b) There is a time delay between the fall of the wolf population and the rise of the moose population.
   (c) Both wolf and moose populations bottomed out in 2009.
   (d) There is an indirect relationship. As moose populations rise, wolf populations also rise.

8. If a population of 10,000 has 300 births, 200 deaths, 50 immigrants and 60 emigrants, what is the populating growth rate?
   (a) 0.9 percent
   (b) 9 percent
   (c) 24 percent
   (d) 90 percent

9. If a country's population growth rate is 5 percent, what is the country's doubling time?
   (a) 5 years
   (b) 14 years
   (c) 42 years
   (d) 72 years

Use the following figure to answer question 10.

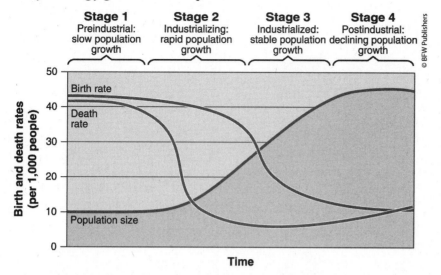

10. Which combination of factors leads to a dramatic increase in total population during Stage 2: Industrializing of the demographic transition model?
    (a) Birth rate falls dramatically due to prevalence of vaccination programs.
    (b) Dramatic increase in death rates due to lack of access to clean water and health care.
    (c) Dramatic rise in total population due to high rates of immigration.
    (d) large difference between birth rates and death rates due to improved technology and access to health care.

11. Which of the following accurately shows how to find the doubling time for Country X given the following information. Country X has a total population of 345,000 individuals. There are 7,500 births, 6,5000 deaths. Net immigration is 3,000 individuals.

    (a) $\frac{(7,500+6,500)-300}{345,000} \times 100 = 3.188$
    $70 \div 0.03188$

    (b) $\frac{(7,500+300)-6,500}{345,000} \times 100 = 1.159$
    $70 \div 1.159$

    (c) $\frac{(7,500+300)-6,500}{345,000} \times 100 = 1.159$
    $70 \div 0.019$

    (d) $\frac{7,500-(6,500+3,000)}{345,000} \times 100 = 1.159$
    $70 \div 0.580$

12. What are two reasons for the rapid growth of the human population over the past 8,000 years?
    (a) increase in warfare
    (b) increases in access to vaccination programs and family planning
    (c) the advent of agriculture and access to technology
    (d) lack of reliable birth control and increasing urban poverty

Use the following graph to answer questions 13 and 14.

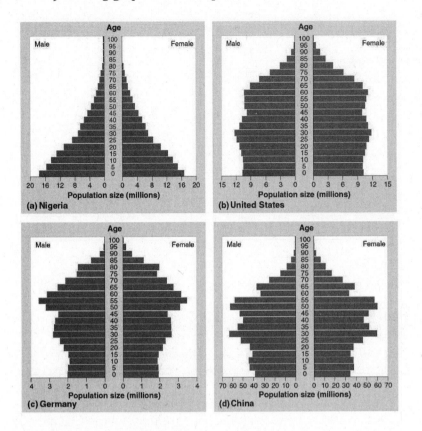

Data from: https://www. *census.gov/population/international/data/idb.*

13. Which country's age structure diagram shows the most stability?
    (a) Nigeria
    (b) China
    (c) Germany
    (d) The United States

14. Examining Nigeria's population structure, we see that
    (a) it is growing rapidly.
    (b) it is industrialized.
    (c) its population momentum is slowing down.
    (d) it has a large immigration rate.

Use the following graph to answer questions 15 and 16.

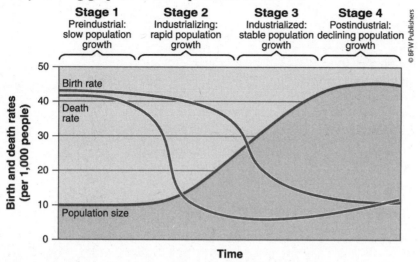

15. When is the population size both large and stable?
    (a) Stage 1
    (b) Stage 2
    (c) Stage 3
    (d) Stage 4

16. Which of the following best describes why the population is stable in Stage 1?
    (a) Birth rate is significantly higher than death rate.
    (b) Both birth and death rates are low, causing the population to decline.
    (c) Birth rates and death rates are relatively equal, causing there to be little changes to the population.
    (d) Birth rates have a direct relationship with death rates and when one increases, the other decreases.

17. When is the population growing most rapidly?
    (a) Between Stages 1 and 2
    (b) Between Stages 2 and 3
    (c) Between Stages 3 and 4
    (d) Following Stage 4

18. If a developing nation quickly reduces its growth rate to 0 percent, its population would
    (a) decrease rapidly.
    (b) decrease slowly.
    (c) level off.
    (d) continue growing for many years then level off.

19. A country with a large, relatively affluent population that is technologically advanced will most likely
    (a) grow exponentially.
    (b) have a large environmental impact.
    (c) have a low GDP.
    (d) have a high emigration rate.

20. If a country has a negative net migration rate and low total fertility rate, what can we say about its population?
    (a) Its population is increasing.
    (b) Its population is falling.
    (c) Its population might continue to increase but over a longer time, it will fall.
    (d) Its population will fall but then rebound and increase.

21. At present, the size of Earth's human population is closest to
    (a) 300 million.
    (b) 1 billion.
    (c) 3.5 billion.
    (d) 8 billion.

Use the following figure to answer question 22.

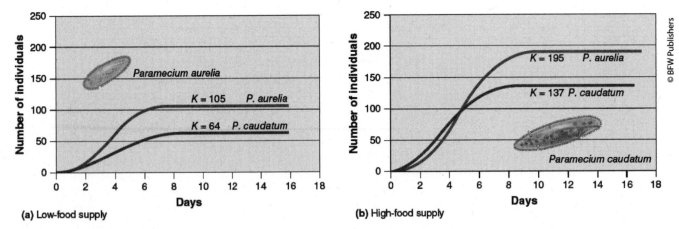

(a) Low-food supply   (b) High-food supply

22. Which of the statements below is supported by the graphs?
    (a) P. aurelia was generally able to outcompete P. caudatum following Day 5 in high-food supply environments.
    (b) P. aurelia always had larger populations than P. caudatum.
    (c) In both low-food and high food supplies P. caudatum was able to outcompete P. aurelia.
    (d) Population sizes were larger in low-food environments than in high food environments.

Use the following figure to answer question 23.

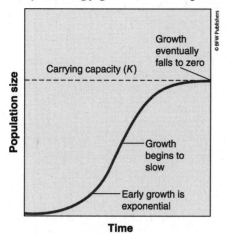

23. Which statement best supports why the population size eventually evens out at the carrying capacity (K)?
    (a) Density independent factors keep the population from overshooting
    (b) Density dependent factors limit population growth as there are insufficient resources.
    (c) There are ample resources for the population to grow at its intrinsic rate of increase.
    (d) High competition for resources keeps the total population low, below carrying capacity.

Use the following figure to answer question 24.

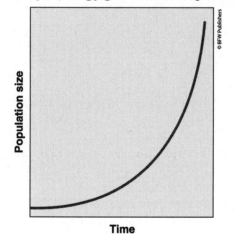

24. Exponential growth models are most likely to be experienced by which of the following populations?
    (a) *K*-selected species living in a well-established area.
    (b) *r*-selected species living in a new environment.
    (c) *K*-selected species that are living in competitive environments.
    (d) *r*-selected species living in competitive environments.

Use the following figure to answer question 25.

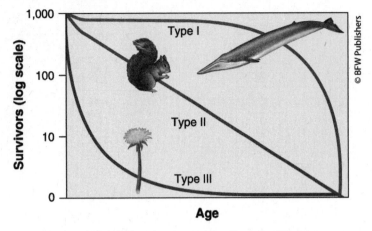

25. Which of the following best describes the life history strategy of Type I species?
    (a) Parents expend much energy on releasing eggs but provide little parental care and therefore offspring typically die young.
    (b) Parents provide some parental care and there is equal chance of death throughout the individual's lifespan.
    (c) Parents expend a lot of energy providing parental care to offspring which generally survive.
    (d) Parents expend little energy or parental care to offspring which typically live a long time.

26. Which of the following correctly describes characteristics of an *r*-selected species?
    (a) typically live long time
    (b) produce many offspring that are typically small
    (c) populations are regulated by density dependent factors
    (d) typically provide parental care

27. Why are *K*-selected species more likely to be affected by invasive species than *r*-selected species?
    (a) *r*-selected species may utilize the same resources as an invasive species, but because their population doesn't grow quickly, they cannot compete.
    (b) *K*-selected species are able to outcompete and grow more rapidly than an *r*-selected species, therefore using more resources.
    (c) *r*- selected species grow slowly and are unable to catch up to *K*-selected species.
    (d) *K*-selected species may utilize the same resources as an invasive species, but their population may be unable to grow and exploit the resource like an invasive species can.

UNIT 4

Earth Systems and Resources

Unit Summary

Earth systems and resources connect abiotic, non-living, ecosystem factors to the biotic, or living portion of the ecosystems. This unit begins with the big movements of Earth's driving mechanisms through plate tectonics, providing a discussion of causes and effects, such as mountains, earthquakes, and volcanoes. Continuing with Earth's processes we then focus on the formation of soil, its properties and layers, along with identifying how to minimize soil erosion. Soils guide us to watersheds and how the water flows, holds onto the water, and the impacts from the human processes. After that the unit examines Earth's atmosphere, global wind patterns, and ocean circulations. The interconnectedness of these topics helps build an understanding of our climate. The final module considers Earth's geography and climate, and the phenomena of El Niño, and La Niña. Earth's systems and resources are hard to place into one simple category, yet they interact as one combined unit that works together to create the planet we inhabit.

MODULES IN THIS UNIT

Module 19: Plate Tectonics
Module 20: Soil Formation, Erosion, Composition, and Its Properties
Module 21: Watersheds
Module 22 Earth's Atmosphere, Global Wind Patterns, Solar Radiation, and Earth's Seasons
Module 23: Earth's Geography and Climate, El Niño, and La Niña

Unit Opening Case: *The Eruption of Mount Saint Helens: 40 Years Later*
The opening case study introduces you to Earth's systems through the volcanic eruption of Mount St. Helens on May 18, 1980 in Washington state. The events of the 1980 eruption provide a glimpse of the devastation caused to the mountain landscape, trees and other vegetation, and animals and the impact to the bodies of water. Scientists have used the eruption to study the succession of the area, the plant and animal interactions, and working to understand volcanoes around the world.

Do the Math
This unit contains the following "Do the Math" box to help prepare you for calculation questions you might encounter on the exam.
- "Plate Movement" (page 235)

To make sure you understand the concepts and techniques presented in this box, do the practice problems presented in the text as well as the additional "Practice the Math" problems that appear in Module 19 of this study guide.

MODULE 19: Plate Tectonics

Before You Read the Module

Focus on Learning Goals

Use the module learning goals to guide your reading. On a separate piece of paper, write down each goal and take notes to help you meet each learning goal. After studying this module, you should be able to:

- 19-1 describe the layers of Earth.
- 19-2 explain how tectonic plate movement produces divergent, convergent, and transform boundaries.
- 19-3 identify how the global distribution of plate boundaries predicts the locations of geologic events.

Key Terms

Core	Earthquake	Convergent boundary
Mantle	Hot spot	Subduction
Magma	Volcano	Island arc
Asthenosphere	Tsunami	Collision zone
Lithosphere	Divergent boundary	Transform boundary
Plate tectonics	Seafloor spreading	Fault

While You Read the Module

Answer the following questions as you read. Use a separate sheet of paper if necessary.

Unit Opening Case: The Eruption of Mount Saint Helens: 40 Years Later

1. Where is Mount Saint Helens located?

2. Describe the explosive event that took place May 18, 1980.

3. Identify the two signs that indicated the mountain was active before the eruption.

4. Explain the how the eruption changed the mountain.

5. Explain how the eruption impacted nearby bodies of water.

6. Explain why the lake needed an outlet tunnel.

7. List how succession started the recovery of the landscape, including the role of pocket gophers.

8. Why did researchers leave the dead trees lying on the ground?

9. Describe the forests today.

Module 19: Plate Tectonics

Earth is comprised of layers from the core to the tectonic plates on the surface

10. How old is Earth?

11. Describe what happened to the elements as Earth cooled.

12. Finish this statement: "In other words, the elements and minerals that were present when the planet formed—and which are distributed. . ."

13. Figure 19.1: Identify Earth's main four layers.

14. Define core.

15. Define mantle.

16. Define magma.

17. Define asthenosphere.

18. Define lithosphere.

Tectonic plate movement creates major geographic features on Earth

The Theory of Plate Tectonics

19. Describe Pangaea.

20. List the evidence Alfred Wegener collected to support the single landmass hypothesis.

21. Figure 19.2: Describe the map locations of the Mesosaurus fossil. Explain how this fossil is an example for Wegener's hypothesis of one land mass.

22. Explain the discovery of Marie Tharp.

23. Define plate tectonics.

Earthquakes and Volcanoes

24. List the evidence that plates are in motion.

25. Define an earthquake.

26. Describe how much plate movement occurs along a fault after an earthquake.

27. Describe the Richter scale.

28. Describe the difference between earthquakes of magnitudes of 7.0 and 6.0, and 6.0 and 4.0.

29. Define hot spot.

30. Define volcano.

31. Figure 19.3: Describe the Pacific Plate hot spot.

32. Define tsunami.

Divergent, Convergent, and Transform Boundaries

33. Identify and describe the two categories of plates.

34. Define divergent plate boundary.

35. Define seafloor spreading.

36. Define convergent plate boundary.

37. Define subduction.

38. Define island arc.

39. Figure 19.4: Describe what drives plate movement.

40. Figure 19.4: Describe where new crust is formed.

41. Figure 19.4: Describe where old oceanic crust is subducted.

42. Define collision zone.

43. Figure 19.5: Identify the type of boundary and feature shown in the figure.

44. Figure 19.6: What direction is the North American plate moving?

45. Figure 19.6: Explain why Earth neither shrinks nor expands in size but stays a constant size.

46. Define transform boundary.

47. Define fault.

48. Figure 19.7: Name and identify the kind of fault in the diagram. Note the direction of movement for the plates.

49. Figure 19.8: List the three types of boundaries, and draw arrows showing directional movement for each side of the lithosphere.

Consequences of Plate Movement on Biodiversity

50. Describe how the separation of continents can cause species to change.

Plate boundaries determine the location of many volcanoes, island arcs, earthquakes, hot spots, and faults

51. Figure 19.9: Describe the "Ring of Fire."

Environmental and Human Impacts of Earthquakes and Volcanoes

52. List the damaging effects of a moderate (5.0-5.9) earthquake.

53. Describe the devastation that can occur from an earthquake in highly populated areas.

54. Describe how nuclear power plants are an area of concern in earthquake prone areas.

55. List the environmental impacts associated with volcanic eruptions.

Practice the Math: Plate Movement

Read "Do the Math: Plate Movement," on page 235. Try "Your Turn." For more math practice, do the following problem. Remember to show your work. Use a separate sheet of paper if necessary.

New Hanover is approximately 500 km north of Shaky Acres. If the plate under Shaky Acres is moving at 15mm per year toward New Hanover, how long will it take for Shaky Acres to be located next to New Hanover? [Formula: time = distance ÷ rate]

After You Read the Module

Review Key Terms
Match the key terms on the left with the definitions on the right.

_____ 1. Core

_____ 2. Mantle

_____ 3. Magma

_____ 4. Asthenosphere

_____ 5. Lithosphere

_____ 6. Plate tectonics

_____ 7. Earthquake

_____ 8. Hot spot

_____ 9. Volcano

_____ 10. Tsunami

_____ 11. Divergent boundary

_____ 12. Seafloor spreading

_____ 13. Convergent boundary

_____ 14. Subduction

_____ 15. Island arc

_____ 16. Collision zone

_____ 17. Transform boundary

_____ 18. Fault

a. The outermost layer of Earth, including the solid upper mantle and crust.

b. Sudden movement in Earth's crust caused by release of potential energy from the movement of tectonic plates.

c. A chain of islands formed by volcanoes as a result of two tectonic plates coming together and experiencing subduction.

d. The innermost zone of Earth's interior, composed mostly of iron and nickel. It includes a liquid outer layer and a solid inner layer.

e. An area where tectonic plates move sideways past each other.

f. An area below the ocean where tectonic plates move away from each other.

g. In geology, a place where molten material from Earth's mantle reaches the lithosphere.

h. A series of waves in the ocean caused by seismic activity or an undersea volcano that causes a massive displacement of water.

i. A vent in the surface of Earth that emits ash, gases, or molten lava.

j. Caused by a divergent boundary, in which rising magma forms new oceanic crust on the seafloor at the boundaries between those plates.

k. The layer of Earth above the core, containing magma, the asthenosphere, and the solid upper mantle.

l. Caused by a divergent boundary, in which rising magma forms new oceanic crust on the seafloor at the boundaries between those plates.

m. The layer of Earth located in the outer part of the mantle, composed of semi-molten rock.

n. An area where two continental plates are pushed together and the colliding forces push up the crust to form a mountain range.

o. The process in which the edge of an oceanic plate moves downward beneath the continental plate and is pushed toward the center of Earth.

p. A fracture in rock caused by movement in Earth's crust.

q. Molten rock.

r. The theory that the lithosphere of Earth is divided into plates, most of which are in constant motion.

MODULE 20: Soil Formation, Erosion, Composition, and Its Properties

Before You Read the Module

Focus on Learning Goals

Use the module learning goals to guide your reading. On a separate piece of paper, write down each goal and take notes to help you meet each learning goal. After studying this module, you should be able to:

- 20-1 describe the formation of igneous, sedimentary, and metamorphic rocks.
- 20-2 identify the processes that break down rocks.
- 20-3 explain how soils are formed.
- 20-4 describe the causes of soil erosion.
- 20-5 identify the properties that affect soil productivity.

Key Terms

Igneous rock	Erosion	B horizon
Sedimentary rock	Parent material	C horizon
Metamorphic rock	Horizon	Porosity
Rock cycle	O horizon	Water holding capacity
Physical weathering	Humus	Permeability
Chemical weathering	A horizon (Topsoil)	Cation exchange capacity (CEC)
Acid precipitation (Acid rain)	E horizon	Base saturation

While You Read the Module

Answer the following questions as you read. Use a separate sheet of paper if necessary.

Module 20: Soil Formation, Erosion, Composition, and Its Properties

The three major types of rock are formed by different combinations of heat and pressure

1. List the three major ways in which rocks we see at Earth's surface form.

2. Figure 20.1: List the three types of rocks

Igneous Rocks

3. Define igneous rock.

4. Describe an igneous vein and its significance.

Sedimentary Rocks

    5. Define sedimentary rocks.

    6. List the environments in which sedimentary rocks form.

Metamorphic Rocks

    7. Define metamorphic rocks.

    8. Describe where the pressure of metamorphic rocks originates.

The Rock Cycle

    9. Define rock cycle.

    10. What part of the rock cycle are environmental scientists most interested in?

    11. Figure 20.2: What are the processes that occur at Earth's surface before a rock becomes sedimentary rock?

Rock is broken down by weathering and erosion

Weathering

    12. Describe when weathering occurs.

    13. Define physical weathering.

    14. Figure 20.3: Describe the two types of physical weathering in the photographs.

    15. Describe why physical weathering increases the rate of chemical weathering.

16. Define chemical weathering.

17. Describe how the chemical weathering of feldspar aids in plant growth.

18. Figure 20.4: Describe the two examples of chemical weathering.

19. Explain how carbon dioxide can lead to chemical weathering.

20. Explain how human activities can cause rain and snow to contain high amounts of sulfuric and nitric acid.

21. Define acid precipitation (acid rain).

22. Explain how knowing rates of weathering can assist researchers.

Erosion

23. Define erosion.

24. Describe the two mechanisms of erosion.

25. Describe deposition.

Soils form as a result of parent material, climate, topography, organisms, and time

26. Figure 20.6: List the ecosystem services of soil.

The Formation of Soil

27. Figure 20.7: Describe what happens to the depth of soil through time.

28. Figure 20.7: Think back to Module 14 and identify the type of succession shown in the first diagram of immature soil and parent rock.

Parent Material

29. Define parent material.

30. Describe the difference between a parent material that is nutrient poor and one that is nutrient rich.

Climate

31. Explain how climate influences soil formation.

Topography

32. Describe topography.

Organisms

33. Explain how organisms influence soil formation.

Time

34. Describe the time it takes for soils to develop.

35. Why are grassland soils considered deep and fertile?

Soil Horizons

36. Define horizon.

37. Define O horizon.

38. Define humus.

39. Define A horizon (topsoil).

40. Define E horizon.

41. Define B horizon.

42. Define C horizon.

43. Figure 20.8: Identify what soil horizon composition depends on.

Soil erodes by wind and water, as well as increases in human activities

44. Explain what soil degradation is and give an example.

Erosion by Water

45. Explain how the topsoil is eroded after vegetation is removed.

46. Figure 20.9a: Describe the large-scale erosion in the photograph.

Erosion by Wind

47. Describe the negative impact of the 1920s and 1930s native grasslands conversion to wheat fields.

48. Identify the causes of "the Dust Bowl."

Soils have different physical, chemical, and biological properties that affect productivity

Physical Properties of Soil

49. Describe the physical properties of soil.

50. Define porosity.

51. Explain how texture of soil is determined.

52. Figure 20.11: Determine the texture of soil: 60 percent sand, 10 percent silt, 30 percent clay.

53. Figure 20.12: Explain how to use a jar with soil and water to determine the type of soil.

54. Define water holding capacity of soil.

55. Describe the permeability of soil.

56. Explain the benefits of a soil with a high percentage of sand.

57. Explain the disadvantages of a soil with a higher proportion of clay.

58. Identify and describe the best agriculture soil.

59. Describe how sandy soils can lead to ground water contamination.

Chemical Properties of Soil

60. Identify and describe the particles that contribute the most to soil chemical properties.

61. Define cation exchange capacity (CEC).

62. Identify the importance of soils with high cation exchange capacity (CEC).

63. List the soil bases.

64. Describe the importance of soil bases.

65. Define base saturation.

66. Describe how cation exchange capacity and base saturation are important to overall ecosystem productivity.

Biological Properties of Soil

67. List the organisms that populate soils and describe their activity.

Review Key Terms

Match the key terms on the left with the definitions on the right.

_____ 1. Igneous rock

_____ 2. Sedimentary rock

_____ 3. Metamorphic rock

_____ 4. Rock cycle

_____ 5. Physical weathering

_____ 6. Chemical weathering

_____ 7. Acid precipitation (Acid rain)

_____ 8. Erosion

_____ 9. Parent material

_____ 10. Horizon

_____ 11. O horizon

_____ 12. Humus

_____ 13. A horizon (Topsoil)

_____ 14. E horizon

_____ 15. B horizon

_____ 16. C horizon

_____ 17. Porosity

_____ 18. Water holding capacity

_____ 19. Permeability

_____ 20. Cation exchange capacity (CEC)

_____ 21. Base saturation

a. The geologic cycle governing the constant formation, alteration, and destruction of rock material that results from tectonics, weathering, and erosion, among other processes.

b. Commonly known as subsoil, a soil horizon is composed primarily of mineral material with very little organic matter.

c. Precipitation high in sulfuric acid and nitric acid.

d. Frequently the top layer of soil, a zone of organic material and minerals that have been mixed together.

e. The mechanical breakdown of rocks and minerals.

f. Rock that forms when sediments such as muds, sands, or gravels are compressed by overlying sediments.

g. Rock formed directly from magma.

h. The size of the air spaces between particles.

i. The underlying rock material from which the inorganic components of a soil are derived.

j. The physical removal of rock fragments from a landscape or ecosystem.

k. The least-weathered soil horizon, which always occurs beneath the B horizon and is similar to the parent material.

l. The amount of water a soil can hold against the draining force of gravity.

m. Rock that forms when sedimentary rock, igneous rock, or other metamorphic rock is subjected to high temperature and pressure.

n. A zone of leaching, or eluviation, found in some acidic soils under the O horizon or, less often, the A horizon.

o. A horizontal layer in a soil defined by distinctive physical features such as color and texture.

p. The ability of a particular soil to adsorb and release cations.

q. The proportion of soil bases to soil acids, expressed as a percentage.

r. The organic horizon at the surface of many soils, composed of organic detritus in various stages of decomposition.

s. The breakdown of rocks and minerals by chemical reactions, the dissolving of chemical elements from rocks, or both these processes.

t. The most fully decomposed organic matter in the lowest section of the O horizon.

u. The ability of water to move through the soil.

MODULE 21: Watersheds

Before You Read the Module

Focus on Learning Goals
Use the module learning goals to guide your reading. On a separate piece of paper, write down each goal and take notes to help you meet each learning goal. After studying this module, you should be able to:
- 21-1 describe the characteristics of a watershed.
- 21-2 identify the impacts humans have on watersheds.

Key Term

Watershed

While You Read the Module
Answer the following questions as you read. Use a separate sheet of paper if necessary.

Module 21: Watersheds

1. Define watershed.

2. Figure 21.1: Look at the diagram and describe where the water drains.

Watersheds are characterized by their area, length, slope, soils, and vegetation

Area and Length

3. How much water does the Mississippi River drain from the United States?

4. Explain how a watershed is measured.

Slope

5. Describe the slope of a watershed.

6. How is erosion impacted by gentle slopes and steep slopes?

Soil Type

7. What is the water permeability difference between sandy soils and clay soils?

8. Describe the negative effect of too much silt and clay in bodies of water.

Vegetation Type

9. List the key roles of vegetation in a watershed.

Humans impact watersheds by altering water flow and inputting excess nutrients and soil

The Hubbard Brook Watersheds

10. Explain what researchers at Hubbard Brook determined about watersheds affected by clear-cutting.

11. Explain what scientists discovered about nitrates when ecological succession was allowed to happen.

The Chesapeake Bay Watershed

12. Describe the location of the Chesapeake Bay watershed.

13. Identify where the water originates from and the consequences of receiving water from this large watershed.

14. List the ecosystem services of the Chesapeake Bay.

15. List the contaminants, nutrients, and sediments of the bay and their origin.

16. Identify the consequence of nutrients flowing into the bay.

17. Explain how suspended sediments in the water affect the grass, fish, and crabs in the bay.

18. Describe the gains of the Chesapeake Bay Action Plan since 2000.

MODULE 22: Earth's Atmosphere, Global Wind Patterns, Solar Radiation, and Earth's Seasons

Before You Read the Module

Focus on Learning Goals
Use the module learning goals to guide your reading. On a separate piece of paper, write down each goal and take notes to help you meet each learning goal. After studying this module, you should be able to:
- 22-1 explain why the amount of solar radiation varies with latitude and the seasons.
- 22-2 identify the major gases and layers in Earth's atmosphere.
- 22-3 describe how the properties of air determine patterns of air circulation.
- 22-4 explain what drives atmospheric convection currents.
- 22-5 explain how the Coriolis effect alters global wind directions.

Key Terms

Insolation	Exosphere	Hadley cell
Albedo	Saturation point	Intertropical convergence
Troposphere	Adiabatic cooling	zone (ITCZ)
Stratosphere	Adiabatic heating	Polar cell
Ozone	Latent heat release	Ferrell cell
Mesosphere	Atmospheric convection	Coriolis effect
Thermosphere	current	

While You Read the Module

Answer the following questions as you read. Use a separate sheet of paper if necessary.

Module 22: Earth's Atmosphere, Global Wind Patterns, Solar Radiation, and Earth's Seasons

1. Define insolation.

2. Explain how solar radiation and the properties of air cause changes around Earth.

Solar radiation varies due to Earth's curvature and tilted axis

Effects of Latitude

3. Identify the three primary causes of the uneven warming patterns of Earth.

4. Figure 22.1: Describe the significance of the two angles and the associated lines shown in the bottom left inset diagram.

5. Why does the length of the Sun's rays cause unequal heating of Earth's surface?

6. Describe how the amount of surface area leads to uneven warming of Earth.

7. Define albedo.

8. Describe the difference between higher and lower albedo and give an example for each.

9. Figure 22.2: List the highest and lowest albedos in the figure and identify the percentage for each. What do you notice about the color of each item?

Effects of Seasons

10. What causes Earth's seasons?

11. Figure 22.3: What hemisphere does this diagram represent?

12. Figure 22.3: How do the seasons differ between the Northern and Southern Hemispheres?

13. When do the Sun's rays strike the equator directly?

14. Describe the June solstice.

15. Describe the December solstice.

The atmosphere is dominated by nitrogen and oxygen gases

16. Figure 22.4: List the two gases that make up 99 percent of the atmosphere. Note the percentage of each.

17. Describe how a small amount of greenhouse gases can have a big effect on Earth's temperature.

The layers of the atmosphere differ in mass, pressure, and temperature

18. Explain what layers of the atmosphere would have the greater mass and how this affects air pressure.

19. Figure 22.5: List Earth's five atmospheric layers.

20. Define troposphere.

21. List the characteristics of the troposphere.

22. Define stratosphere.

23. Define ozone.

24. What is the purpose of the ozone layer?

25. Why is ultraviolet (UV) radiation harmful?

26. Figure 22.5: Why is the upper stratosphere warmer?

27. Define the mesosphere.

28. Define the thermosphere.

29. Describe how the thermosphere is useful.

30. Describe the northern lights (aurora borealis) and the southern lights (aurora australis).

31. Define the exosphere.

32. What can be found in the exosphere?

33. Figure 22.5: Describe the trend of the plotted temperature line compared to increasing altitude for each atmospheric layer.

Air circulates in the atmosphere as a result of changing density, water vapor capacity, and temperature

34. What are the four properties of air that determine circulation?

35. Describe air density.

36. How do density and temperature determine air movement?

37. How does the ability of air to hold water change with temperature?

38. Define saturation point.

39. What happens to water vapor in the air if the air temperature drops?

40. Figure 22.7: What would happen to the air vapor if the temperature dropped from 30°C to 10°C?

41. Describe how pressure changes in the atmosphere.

42. Define adiabatic cooling.

43. Define adiabatic heating.

44. Define latent heat release.

45. Describe the effect on the air after latent heat release has occurred.

Atmospheric convection currents are driven by solar radiation at the equator

46. Define atmospheric convection currents.

47. Figure 22.8: List the steps that occur to produce atmospheric currents. Start at the bottom of the convection cell, "hot, dry air picks up moisture...,"

48. Define Hadley cells.

49. What causes Hadley cells?

50. Define intertropical convergence zone (ITCZ).

51. Explain why the latitudes of the intertropical convergence zone (ITCZ) are not fixed.

52. Define polar cells.

53. Define Ferrell cells.

54. How are Ferrell cells helpful in distributing air in the atmosphere?

Earth's rotation causes the Coriolis effect, which deflects global wind patterns

55. Define Coriolis effect.

56. Figure 22.10a: Looking at the North Pole, what direction does the Earth appear to move, clockwise or counterclockwise?

57. Figure 22.10a: Describe the direction that the ball travels when thrown from the North Pole towards the equator.

58. Figure 22.10b: Describe the direction that the ball travels when thrown from the South Pole towards the equator.

59. Figure 22.10b: What will happen to the direction of the ball if it is thrown from the equator to either pole, North or South?

60. Describe why the deflection of objects moving north or south occurs.

61. Figure 22.11: List the speeds of Earth's rotation at each of the differing latitudes.

62. Why is there a difference in the speed of Earth's rotation at different latitudes?

63. Describe how the atmospheric convection currents and Coriolis effect work together.

64. Figure 22.12: List each of the prevailing wind patterns and their direction of movement in each latitude section.

After You Read the Module

Review Key Terms
Match the key terms on the left with the definitions on the right.

_____ 1. Insolation

_____ 2. Albedo

_____ 3. Troposphere

_____ 4. Stratosphere

_____ 5. Ozone

_____ 6. Mesosphere

_____ 7. Thermosphere

_____ 8. Exosphere

_____ 9. Saturation point

_____ 10. Adiabatic cooling

_____ 11. Adiabatic heating

_____ 12. Latent heat release

_____ 13. Atmospheric convection current

_____ 14. Hadley cell

_____ 15. Intertropical convergence zone (ITCZ)

_____ 16. Polar cell

_____ 17. Ferrell cell

_____ 18. Coriolis effect

a. The layer of the atmosphere above the stratosphere, extending roughly 50 to 85 km (31–53 miles) above the surface of Earth.

b. A layer of the atmosphere closest to the surface of Earth, extending up to approximately 16 km (10 miles).

c. The deflection of an object's path due to the rotation of Earth.

d. The layer of the atmosphere above the mesosphere, extending 85 to 600 km (53–375 miles) above the surface of Earth.

e. The cooling effect of reduced pressure on air as it rises higher in the atmosphere and expands.

f. Global patterns of air movement that are initiated by the unequal heating of Earth.

g. Incoming solar radiation, which is the main source of energy on Earth.

h. The release of energy when water vapor in the atmosphere condenses into liquid water.

i. A convection current in the atmosphere that cycles between the equator and 30° N and 30° S.

j. A convection current in the atmosphere, formed by air that rises at 60° N and 60° S and sinks at the poles, 90° N and 90° S.

k. A pale blue gas composed of molecules made up of three oxygen atoms (O_3).

l. The percentage of incoming sunlight reflected from a surface.

m. The maximum amount of water vapor in the air at a given temperature.

n. The latitude that receives the most intense sunlight, which causes the ascending branches of the two Hadley cells to converge.

o. The outermost layer of the atmosphere, which extends from 600 to 10,000 km (375– 6,200 miles) above the surface of Earth.

p. The heating effect of increased pressure on air as it sinks toward the surface of Earth and decreases in volume.

q. The layer of the atmosphere above the troposphere, extending roughly 16 to 50 km (10–31 miles) above the surface of Earth.

r. A convection current in the atmosphere that lies between Hadley cells and polar cells.

MODULE 23: Earth's Geography and Climate, El Niño, and La Niña

Before You Read the Module

Focus on Learning Goals
Use the module learning goals to guide your reading. On a separate piece of paper, write down each goal and take notes to help you meet each learning goal. After studying this module, you should be able to:
- 23-1 describe how ocean circulation affects weather and global climates.
- 23-2 explain how mountains interrupt ocean airflow to cause rain shadows.
- 23-3 identify the causes and consequence of El Niño and La Niña events.

Key Terms

Gyre	Rain shadow	La Niña
Upwelling	El Niño–Southern Oscillation	Carbon sequestration
Thermohaline circulation	(ENSO)	Aqueducts

While You Read the Module
Answer the following questions as you read. Use a separate sheet of paper if necessary.

Module 23: Earth's Geography and Climate, El Niño, and La Niña

Ocean currents are driven by unequal heating, gravity, wind, salinity, and the location of continents

1. Why is the flow of ocean water an important factor for global climate?

Unequal Heating and Gravity

2. What drives ocean currents?

3. What happens to water that receives the most direct sunlight?

Wind and the Coriolis Effect

4. Define gyres.

5. Figure 23.1: Find the North Pacific Current in the Northern Hemisphere and describe the movement of the gyre.

6. Figure 23.1: Find the South Equatorial Current in the Southern Hemisphere and describe the movement of the gyre.

7. Figure 23.1: What drives each of the five large gyres?

8. Explain how the California Current causes cooler temperatures along the coastal regions.

Upwelling

9. Define upwelling.

10. Describe the effects of an upwelling on fish populations.

Deep Ocean Currents

11. Define thermohaline circulation.

12. What is significant about thermohaline circulation?

13. Describe the processes that drive thermohaline circulation.

14. How long does the complete process of thermohaline circulation take?

15. Why are scientists concerned that global warming could affect thermohaline circulation?

Rain shadows cause mountains to be dry on one side

16. Figure 23.3: Describe windward and leeward sides of the mountains.

17. Define rain shadow.

18. Figure 23.4: Identify the different biomes of the Cascade Mountain range in Washington in relation to the rain shadow effect.

The El Niño–Southern Oscillation is caused by a shift in prevailing winds and ocean currents that alters global weather

19. How often does an El Niño-Southern Oscillation typically occur and when?

20. Define El Niño-Southern Oscillation (ENSO).

21. List the steps to the formation of an El Niño-Southern Oscillation.

22. List the global consequences of an El Niño-Southern Oscillation.

23. Define La Niña.

Visual Representation 4: Earth System in Iceland

24. Explain the sources of water for Lake Thingvallavatn.

25. Identify the three environments of soil and describe their formation.

26. Name the geologic boundary and tectonic plates affecting Iceland. Describe the movement of each plate.

27. What active geologic features are common in Iceland?

28. Identify the ocean currents that affect Iceland. Describe how each current impacts the climate on the Island.

29. Identify the amount of solar radiation that is received in Iceland. Provide an explanation for your answer.

Pursuing Environmental Solutions: Sequestering Carbon to Reduce Global Warming

30. Define carbon sequestration.

31. Identify the three approaches to carbon sequestration.

32. List the biological approach to management strategies to increase carbon storage in agricultural soils.

33. Describe new technologies to pull CO_2 out of the air.

34. Give an example of CO_2 capture and its use in the industrial and manufacturing processes.

35. When it comes to carbon sequestration, what do the experts agree would be the best option?

Science Applied 4

Can We Resolve the California Water Wars?

36. Describe the amount of agriculture industry California accounts for in the United States.

37. Define aqueducts.

38. List the different groups that require or demand water.

What historical factors have led to California's water wars?

39. Describe the 1935 Central Valley Project.

40. Describe the long-term locked contracts that made agriculture possible.

How might farmers respond to increased water prices?

41. Identify how farmers may respond to water price increases.

42. From the four choices a farmer would face if water prices increased, describe the possible outcomes that could occur.

The motivation provided by drought

43. Describe the term "fallow" with regard to agriculture.

44. List the substantial ways the state of California responded to the drought for the past two decades.

How have the water wars played out in the political arena?

45. List and describe the different political maneuverings with the past water conflicts.

What can be done?

46. Explain what economists argue about users of water.

Review Key Terms

Match the key terms on the left with the definitions on the right.

_____ 1. Gyre

_____ 2. Upwelling

_____ 3. Thermohaline circulation

_____ 4. Rain shadow

_____ 5. El Niño–Southern Oscillation (ENSO)

_____ 6. La Niña

_____ 7. Carbon sequestration

_____ 8. Aqueducts

a. A large-scale pattern of water circulation that moves clockwise in the Northern Hemisphere and counterclockwise in the Southern Hemisphere.

b. A reversal of wind and water currents in the South Pacific.

c. Following an El Niño event, trade winds in the South Pacific reverse strongly, causing regions that were hot and dry to become cooler and wetter.

d. The capture and long-term storage of carbon dioxide from the atmosphere.

e. A region with dry conditions found on the leeward side of a mountain range as a result of humid winds from the ocean causing precipitation on the windward side.

f. The upward movement of ocean water toward the surface as a result of diverging currents.

g. Pipes and canals that move water from where it is abundant to areas where it is scarce.

h. An oceanic circulation pattern that drives the mixing of surface water and deep water.

UNIT 4 Review Exercises

Check Your Understanding

Review "Learning Goals Revisited" on pages 237, 251, 257, 268, and 278 of your textbook. Compare the notes you took while reading each module. Complete these exercises to review the chapter. Use a separate piece of paper if necessary.

1. Describe the plate tectonic boundary types and features and/or effects associated with each one.

2. Explain how weathering is a precursor for the soil development process.

3. List the five factors that determine soil development.

4. List and describe the soil horizons.

5. Describe the factors that comprise a watershed and the effect of each one.

6. Describe the factors that determine insolation, solar radiation, and the heating of Earth.

7. Draw Figure 22.5, page 262, on a separate sheet making sure to label the altitudes on the y axis and the temperatures on the x axis.

8. Describe the rain shadow effect and how precipitation patterns change.

9. Describe the three types of ocean circulations.

10. Describe how an El Niño begins.

Practice for Free-Response Questions

Complete this exercise to build and practice the skills you will need to answer free-response questions on the exam. Use a separate sheet of paper if necessary.

1. Rock Cycle Analysis:

   (a) Explain the rock cycle path for a sedimentary rock geologically changing to an igneous volcanic rock.

   (b) Describe how a convergent boundary leads to the formation of a volcanic igneous rock through subduction.

   (c) Explain how the igneous rock could develop into the abiotic portion of soil.

2. Soil formation is a mixture of biotic and abiotic components.

   (a) Identify and explain one component of how texture influences water movement in soils.

   (b) Describe one advantage of organisms in soil.

3. Explain the difference between the amount of solar radiation at the North Pole compared to the Equator.

Unit 4 Multiple-Choice Review Exam

1. The Earth's plates are in constant motion because
    (a) pressure from heat and gas in Earth's core pushes plates sideways.
    (b) circulation in Earth's mantle causes oceanic plates to spread.
    (c) the density of the lithosphere causes plates to sink.
    (d) hot spots create pressure that shifts plates.

2. Which is associated with seafloor spreading?
    (a) Divergent plate boundaries
    (b) Convergent plate boundaries
    (c) Subduction zones
    (d) Hotspots

3. Measured on the Richter scale, an earthquake with a magnitude of 6.0 is how many times greater than an earthquake with a magnitude of 3.0?
    (a) 10
    (b) 100
    (c) 1,000
    (d) 10,000

4. Worldwide, earthquakes typically occur
    (a) every few years.
    (b) once a year.
    (c) many times each day.
    (d) several times each year.

5. Volcanoes are likely to be found in an area where tectonic plates are
    (a) diverging and moving sideways past each other.
    (b) moving past each other and converging.
    (c) diverging and collision zones.
    (d) converging and diverging.

6. Plate movement occurs in which part of Earth's crust?
    (a) Core
    (b) Upper mantle
    (c) Asthenosphere
    (d) Lithosphere

7. Which processes are involved in the formation of sedimentary rock?
    (a) cooling and crystallization
    (b) uplift and heat
    (c) heat and pressure
    (d) erosion, transport and compression

8. Fossils are most likely found in
    (a) igneous rock.
    (b) metamorphic rock.
    (c) sedimentary rock.
    (d) basalt rock.

9. Acid precipitation can cause
   (a) physical weathering.
   (b) erosion.
   (c) subduction.
   (d) chemical weathering.

10. Which type of soil allows the most water to infiltrate?
    (a) sand
    (b) silt
    (c) humus
    (d) loam

11. Soil found in tropical rain forests is generally
    (a) rich in organic material.
    (b) rich in quartz sand.
    (c) deep and porous.
    (d) nutrient poor.

12. The soil layer most similar to the parent material is the
    (a) O horizon.
    (b) A horizon
    (c) B horizon.
    (d) C horizon.

13. Which item contributes to soil formation?
    (a) machines.
    (b) plowing.
    (c) clearcutting.
    (d) composting.

Use the following figure to answer question 14.

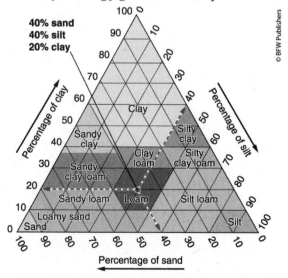

14. Identify the correct percentage of sand, silt, and clay for clay loam.
    (a) sand 20%, silt 30%, clay 50%
    (b) sand 30%, silt 35%, clay 35%
    (c) sand 45%, silt 15%, clay 40%
    (d) sand 40%, silt 40%, clay 20%

15. Which is one way to positively impact watersheds?
    (a) Planting trees.
    (b) Allowing animal waste into the river.
    (c) Spraying your grass for weeds.
    (d) Keeping your crop fields barren until the next planting.

16. The Hubbard Brook experiment showed that
    (a) deforestation increases nutrient runoff.
    (b) river restoration can take my years to complete.
    (c) evapotranspiration increases with more vegetation cover.
    (d) freshwater aquatic ecosystems are often very resilient.

17. The Chesapeake Bay Action Plan outlined a series of goals to reduce impacts and clean up the waters which included
    (a) an increase in nutrients, such as nitrogen.
    (b) an increase to the crab population.
    (c) a decrease the water clarity.
    (d) an increase in sediments in the water.

18. In which layer of Earth's atmosphere does the ozone layer appear?
    (a) troposphere
    (b) stratosphere
    (c) mesosphere
    (d) thermosphere

19. In which layer of Earth's atmosphere does all of our weather occur?
    (a) thermosphere
    (b) exosphere
    (c) troposphere
    (d) stratosphere

20. The northern lights, also known as the aurora borealis, is a glowing light produced by solar radiation energizing gases in the _____ layer of the atmosphere.
    (a) troposphere
    (b) stratosphere
    (c) mesosphere
    (d) thermosphere

21. The prevailing wind systems of the world are produced by
    (a) convection currents and the Coriolis effect.
    (b) convection currents and adiabatic cooling.
    (c) convection currents and the adiabatic heating.
    (d) ocean circulation and the Coriolis effect.

22. During which months is the Sun directly overhead at the equator?
    (a) January and June
    (b) March and September
    (c) December and June
    (d) July and February

23. The El Nino-Southern Oscillation brings what type of weather conditions to the southeastern United States.
    (a) warmer, drier
    (b) warmer, wetter
    (c) cooler, drier
    (d) cooler, wetter

24. Which statement best explains why polar regions are colder than tropical regions?
    (a) Polar regions have lower albedo values.
    (b) Polar regions receive less solar energy per unit of surface area.
    (c) Tropical regions receive less direct sunlight throughout the year.
    (d) Sunlight travels through more atmosphere and loses more energy in tropical regions.

25. Earth's axis of rotation is tilted 23.5°, this is a cause of
    (a) an albedo effect.
    (b) the Hadley cells.
    (c) the seasons.
    (d) the thermohaline circulation.

26. A darker surface on Earth will have _____ albedo compared to a lighter surface.
    (a) lower
    (b) higher
    (c) equal
    (d) no effect

27. The Coriolis effect causes air deflection in the mid-latitudes (between 30° and 60°) of the United States to move the air
    (a) from east to west.
    (b) from west to east.
    (c) from north to south.
    (d) from south to north.

28. The large, scale ocean surface currents are known as:
    (a) upwellings.
    (b) thermohaline circulations.
    (c) Hadley cells.
    (d) gyres.

29. A deep ocean current, that is crucial for moving heat and nutrients around the globe is known as
    (a) upwelling.
    (b) thermohaline circulation.
    (c) Hadley cell.
    (d) gyres.

30. Which divergent plates are responsible for Iceland moving apart 2 cm per year?
    (a) North American Plate and Pacific Plate
    (b) Nazca Plate and South American Plate
    (c) North American Plate and Eurasian Plate
    (d) Eurasian Plate and African Plate

UNIT 5

Land and Water Use

Unit Summary

Now that you have engaged in a thorough understanding of how ecosystems work, we can begin to study how human activities affect ecosystems. This will enable you to understand how changes in human behavior can minimize impacts to ecosystems and the organisms that occupy them. For example, agricultural practices, fishing, and logging have tremendous impacts on the environment. By finding alternative methods of harvesting or utilizing natural resources, we can ensure their availability for years to come. Careful planning for growing cities can also minimize the impacts that humans have on ecosystems and ensure that cities thrive despite larger human populations.

MODULES IN THIS UNIT

Module 24: The Tragedy of the Commons; Clear-cutting
Module 25: The Green Revolution
Module 26: Impacts of Agricultural Practices
Module 27: Irrigation and Pest Control Methods
Module 28: Meat Production Methods and the Impacts of Overfishing
Module 29: Impacts of Mining
Module 30: Impacts of Urbanization and the Methods to Reduce Urban Runoff
Module 31: Ecological Footprints
Module 32: Introduction to Sustainability
Module 33: Integrated Pest Management and Sustainable Agriculture
Module 34: Aquaculture and Sustainable Forestry

Unit Opening Case: *Who Can Use Our Public Lands? And for What Purpose?*

The case study describes several recent conflicts over the use of public lands. Twenty-seven-year-old Tim DeChristopher spent almost two years in prison for bidding on the oil and gas leases that he had no intention of paying. DeChristopher wanted to prevent the land from being used for oil and gas extraction. In another case, Terry Tempest and Brooke Williams attempted to purchase oil and gas leases with the express intention of using the land to develop alternate, cleaner, energy resources such as solar and wind. Their lease application was denied, and they are currently suing the federal government. A final case notes that advocates for the protection of wolves have called for the removal of cattle from public land because Washington state officials have been killing wolves that threaten cattle. Each of these three examples raises questions about the use of public lands and whether or not the citizens of the United States have the right to use public lands for purposes other than those designated by the government.

Do the Math

This unit contains the following "Do the Math" boxes to help prepare you for calculation questions you might encounter on the exam.

- "Changes in the Amount of Protected Lands Over Time" (page 298)
- "Expressing Protected Land Areas as a Percentage" (page 303)
- "Land Needed for Food" (page 314)
- "Calculating Per Capita Water Use" (page 324)
- "What Fraction of Automobiles Are in the Developing World?" (page 343)
- "Rates of Forest Clearing" (page 367)
- "Comparing Carbon Accumulation in Tilled and No-Till Agricultural Soils" (page 375)

To make sure you understand the concepts and techniques presented in these boxes, do the practice problems presented in the text as well as the additional "Practice the Math" problems that appear in Module 24, Module 25, Module 27, Module 29, Module 32, and Module 33 of this study guide.

MODULE 24: The Tragedy of the Commons; Clear-cutting

Before You Read the Module

Focus on Learning Goals

Use the module learning goals to guide your reading. On a separate piece of paper, write down each goal and take notes to help you meet each learning goal. After studying this module, you should be able to:

- 24-1 explain the tragedy of the commons.
- 24-2 describe rangeland and forest land management.
- 24-3 explain the environmental impacts of clear-cutting.

Key Terms

Tragedy of the commons	Forest	Ecologically sustainable forestry
Externality	Clear-cutting	Tree plantation
Rangelands	Selective cutting	Endangered Species Act

While You Read the Module

Answer the following questions as you read. Use a separate sheet of paper if necessary.

Unit Opening Case: Who Can Use Our Public Lands? And for What Purpose?

1. Identify the activities for which the government sells or leases public lands.

2. Describe the efforts of Tim DeChristopher to stop the use of public land to extract development of oil and gas.

3. Explain the reason Terry Tempest and Brooke Williams sued the federal government in 2016.

4. Describe the conflicts between advocates for wolves and ranchers on public lands in Washington.

5. Identify the main questions in the conflict of public land use.

Module 24: The Tragedy of the Commons; Clear-cutting

The tragedy of the commons describes the harm that can occur to public and shared resources

6. List the examples of environmental effects from humans and land use.

The Tragedy of the Commons

    7. How has land been viewed as a common resource in the past?

    8. Define tragedy of the commons.

    9. Figure 24.2: Describe the sheep, farming, and land changes through time in each of the three parts.

   10. Figure 24.2: Describe the tragedy of the commons.

   11. Describe other examples of tragedy of the commons.

Externalities

   12. Define externalities.

   13. Explain an example of a positive externality.

   14. Explain an example of a negative externality.

   15. Describe a result of a negative externality imposed by the sheep farmer shown in Figure 24.2.

   16. Explain how the farmer's actions might change if the farmer is required to pay for use of the land to graze sheep.

   17. Explain why private land is not typically subject to the negative externalities seem in the tragedy of the commons.

Public lands are classified according to their use

18. Figure 24.4 Describe the reasons people have created protected marine areas around the world.

Public Lands in the United States

19. Figure 24.5: Describe the location where most United States federal lands are located.

Land management differs in rangelands and forests

20. Describe the significance of Aldo Leopold's work as a professor, ecologist, writer, and environmentalist.

Rangelands

21. Define rangelands.

22. Identify negative attributes of grazing livestock on rangelands.

Forests

23. Define forests.

24. Describe how national forests work with commercial logging companies.

Clear-cutting can adversely impact soils, streams, and atmospheric carbon dioxide concentrations

25. Define clear-cutting.

26. Provide two advantages for clear cutting.

27. What do foresters do with land after clear-cutting?

28. What kind of species thrive after clear-cutting? How is biodiversity impacted?

Clear-cutting

29. List the possible environmental consequences or costs of clear-cutting.

Selective Cutting and Ecologically Sustainable Forestry

30. Define selective cutting.

31. 26. Figure 24.7(b): Describe the conditions for regrowth of trees in an area that has been subject to selective cutting.

32. Compare and contrast the environmental consequences of selective-cutting and clear-cutting.

33. Define ecologically sustainable forestry.

34. Describe a negative factor related to ecologically sustainable forestry.

35. Figures 24.8 and 24.9: Identify the obvious environmental differences between both photos.

Forest management influences biodiversity in a variety of ways

36. Explain the negative effect of logging on the marbled murrelet.

Reforestation

37. Define tree plantation.

38. Describe a negative consequence of tree plantations.

Federal Regulation of Land Use

39. What guidance about managing public lands do federal regulations give to the U.S. Forest Service?

40. Describe the requirements for environmental impact statements if federal permits or money are used in a project.

41. Define Endangered Species Act.

42. Explain why someone might attend informational sessions about local land use decisions.

Practice the Math: Changes in the Amount of Protected Lands Over Time

Read "Do the Math: Changes in the Amount of Protected Lands Over Time" on page 298. Try "Your Turn." For more math practice, do the following problems. Remember to show your work. Use a separate sheet of paper if necessary.

In Indonesia, both national parks and nature and game reserves are protected land. National parks cover 164,000 km², and the nature and game reserves cover 283,000 km². Indonesia has a total land area of 1.9 million km².

(a) Convert the total amount of protected land and the total landmass of Indonesia to hectares and express in scientific notation.

(b) Identify what percentage of Indonesia's total landmass is protected.

Practice the Math: Expressing Protected Land Areas as a Percentage

Read "Do the Math: Changes in the Amount of Protected Lands Over Time" on page 303. Try "Your Turn." For more math practice, do the following problems. Remember to show your work. Use a separate sheet of paper if necessary.

In the United States, total land cover is 9,490,391 square kilometers; 3,623,150 square kilometers is used as farmland.

(a) Convert to hectares and express in scientific notation for each.

(b) Identify the percentage of farmland in the United States.

After You Read the Module

Review Key Terms
Match the key terms on the left with the definitions on the right.

_____ 1. Tragedy of the commons

_____ 2. Externality

_____ 3. Rangelands

_____ 4. Forest

_____ 5. Clear-cutting

_____ 6. Selective cutting

_____ 7. Ecologically sustainable forestry

_____ 8. Tree plantation

_____ 9. Endangered Species Act

a. A method of harvesting trees that involves removing all or almost all of the trees within an area.

b. The cost or benefit of a good or service that is not included in the purchase price of that good or service, or otherwise accounted for.

c. A 1973 U.S. law designed to protect plant and animal species that are threatened with extinction, and the habitats that support those species.

d. Dry, open grasslands primarily used for grazing cattle.

e. The method of harvesting trees that involves the removal of single trees or a relatively small number of trees from the larger forest.

f. An approach to removing trees from forests in ways that do not unduly affect the viability of other noncommercial tree species.

g. The tendency of a shared, limited resource to become depleted if it is not regulated in some way.

h. Land dominated by trees and other woody vegetation and sometimes used for commercial logging.

i. A large area typically planted with a single fast-growing tree species.

MODULE 25: The Green Revolution

Before You Read the Module

Focus on Learning Goals

Use the module learning goals to guide your reading. On a separate piece of paper, write down each goal and take notes to help you meet each learning goal. After studying this module, you should be able to:

- 25-1 describe changes in agricultural practices that have occurred throughout human civilization.
- 25-2 explain changes that developed as a result of the Green Revolution.
- 25-3 identify the benefits and consequences of Green Revolution innovations.
- 25-4 identify specific benefits and consequences of GMOs.
- 25-5 describe the resulting efficiency and fossil fuel impacts of mechanization.

Key Terms

Subsistence farming
Industrial agriculture (Agribusiness)
Green Revolution
Economies of scale
Organic fertilizer
Synthetic fertilizer (Inorganic fertilizer)
Waterlogging
Salinization

Pesticide
Insecticide
Herbicide
Broad-spectrum pesticide
Selective pesticide (Narrow-spectrum pesticide)
Monocropping
Energy subsidy

While You Read the Module

Answer the following questions as you read. Use a separate sheet of paper if necessary.

Module 25: The Green Revolution

Agriculture has undergone a number of transformations over time

1. Distinguish between agriculture and domestication.

2. Describe the Second Agricultural Revolution.

3. Define subsistence farming.

4. Define industrial agriculture, or agribusiness.

The Green Revolution led to greater food production

5. Define Green Revolution.

6. Describe the contributions of Norman Borlaug.

7. Describe the movement of the Green Revolution from developed countries to developing countries throughout the 20th century.

The Green Revolution led to advances in mechanization, fertilization, irrigation, and the use of pesticides

Mechanization

8. List the work associated with farming.

9. Why are machines used in farming?

10. Define economies of scale.

11. Explain how machines benefit large farms more than small farms.

12. Explain the reason that single crop farms are more efficient than farms that grow multiple crops.

Fertilizers

13. Explain why soil used for industrial agriculture needs large amounts of fertilizers.

14. Define organic fertilizer.

15. Define synthetic fertilizer, or inorganic fertilizers.

16. Identify the advantages of synthetic fertilizers.

17. Identify the adverse effects that synthetic fertilizers have on the environment.

18. Figure 25.3: Identify both a beneficial and a possible harmful effect of synthetic fertilizers shown in the photograph.

Irrigation

19. Describe the benefits of irrigation.

20. Describe the drawbacks of irrigation.

21. Define waterlogging.

22. Define salinization.

Pesticides

23. Define pesticide.

24. Define insecticide.

25. Define herbicide.

26. Define broad-spectrum pesticide and give an example.

27. Define selective pesticide or narrow-spectrum pesticide and give an example.

Monocropping

28. Define monocropping.

29. List dominant monocrops in the United States.

30. Describe the main benefits of monocropping.

31. Describe potential environmental degradation caused by monocropping.

Genetic engineering and the introduction of genetically modified organisms are revolutionizing agriculture

The Benefits of Genetic Engineering

32. Describe how GMOs can increase crop yield and quantity.

33. Explain how scientists have developed golden rice and the possible benefits.

34. Describe the benefit of GMO salmon.

35. Compare GMO crop seeds to conventional crop seeds.

36. Identify how GMOs are cost effective for farmers and consumers.

Concerns About Genetically Modified Organisms

37. List three crops that are typically GMO in the United States and the percent of each that is GMO.

38. Explain what researchers are specifically studying that is a concern for human consumption of GMOs.

39. Explain the possible detrimental effects of GMO crops on biodiversity.

40. Describe the GMO labeling law and its implementation.

41. Explain the status of GMOs in the European Union.

42. Describe the status of GMO animals in the United States.

Mechanization results in the energy subsidy in agriculture

43. Define energy subsidy.

44. Figure 25.9: Explain a reason why hunting and gathering has such a low energy subsidy number.

45. Figure 25.9: Hypothesize why far-offshore fishing has such a high energy subsidy number and explain your reasoning.

46. Figure 25.9: Identify and analyze what you eat on the scale. Describe how you could lower your energy subsidy by eating differently.

47. Explain how a consumer who eats a supermarket diet has a higher energy subsidy.

Practice the Math: Land Needed for Food

Read "Do the Math: Land Needed for Food," on page 314. Try "Your Turn." For more math practice, do the following problem. Remember to show your work. Use a separate sheet of paper if necessary.

On farms in the Midwestern United States, a hectare of land yields roughly 40 bushels of wheat. Each bushel of wheat weighs approximately 27 kilograms. One kilogram of wheat provides 3,500 kilocalories. Assume that a person only eats wheat and must consume 2,000 kcal per day. How much land is required to feed that person for one year?

Review Key Terms

Match the key terms on the left with the definitions on the right.

_____ 1. Subsistence farming

_____ 2. Industrial agriculture (Agribusiness)

_____ 3. Green Revolution

_____ 4. Economies of scale

_____ 5. Organic fertilizer

_____ 6. Synthetic fertilizer (Inorganic fertilizer)

_____ 7. Waterlogging

_____ 8. Salinization

_____ 9. Pesticide

_____ 10. Insecticide

_____ 11. Herbicide

_____ 12. Broad-spectrum pesticide

_____ 13. Selective pesticide (Narrow-spectrum pesticide)

_____ 14. Monocropping

_____ 15. Energy subsidy

a. A pesticide that targets a narrow range of organisms.

b. A form of soil degradation that occurs when the small amount of salts in irrigation water becomes highly concentrated on the soil surface through evaporation.

c. A pesticide that kills many different types of pest.

d. Fertilizer composed of organic matter from plants and animals.

e. A substance, either natural or synthetic, that kills or controls organisms that people consider pests.

f. A pesticide that targets plant species that compete with crops.

g. A shift in agricultural practices in the twentieth century that included new management techniques and mechanization, as well as the triad of fertilization, irrigation, and improved crop varieties, that resulted in increased food output.

h. The fossil fuel energy and human energy input per calorie of food produced.

i. Farming for consumption by the farming family and maybe a few neighbors.

j. An agricultural method that utilizes large plantings of a single species or variety.

k. A pesticide that targets species of insects and other invertebrates that consume crops.

l. Agriculture that applies the techniques of mechanization and standardization to the production of food.

m. The observation that average costs of production fall as output increases.

n. Fertilizer produced commercially, normally with the use of fossil fuels.

o. A form of soil degradation that occurs when soil remains under water for prolonged periods.

MODULE 26: Impacts of Agricultural Practices

Before You Read the Module

Focus on Learning Goals

Use the module learning goals to guide your reading. On a separate piece of paper, write down each goal and take notes to help you meet each learning goal. After studying this module, you should be able to:

- 26-1 explain how plowing and tilling interfere with the natural progression of ecosystems.
- 26-2 identify the consequences of slash and burn farming.
- 26-3 describe the effects of fertilizer use.

Key Terms

Plowing   Tilling   Slash-and-burn agriculture
                                                                                       (Shifting agriculture)

While You Read the Module

Answer the following questions as you read. Use a separate sheet of paper if necessary.

Module 26: Impacts of Agricultural Practices

Many agricultural practices work against succession and other natural processes

1. Describe two ways that agriculture practices work against natural processes.

Plowing and tilling reverse succession and mobilize organic matter

2. Define plowing.

3. Describe benefits of plowing.

4. Define tilling.

5. What is the net result of both plowing and tilling?

6. What are the disadvantages of plowing or tilling?

7. Describe soil compaction.

8. What is the plow layer?

Slash-and-burn farming causes short-term gains and long-term losses

9. Review the soil characteristics of rainforests in Central and South America and other places.

10. Define slash-and-burn agriculture, also known as shifting agriculture.

11. Explain why nutrients are quickly depleted when the slash-and-burn technique is used.

12. Identify the detrimental effects of shifting agriculture on the environment.

13. Figure 26.4: Describe the net destruction of forests.

14. Describe how slash and burn agriculture is a source of CO_2 and particulates.

Fertilizer use affects soil and water

15. Describe the nutrients found in fertilizers and how they are labeled.

16. Compare and contrast synthetic and organic fertilizers.

17. Describe how fertilizer production contributes to global climate change.

18. Provide three advantages of synthetic fertilizers.

19. Describe the disadvantages of synthetic fertilizers

20. Describe what happens when fertilizers runoff into adjacent waterways.

After You Read the Module

Review Key Terms
Match the key terms on the left with the definitions on the right.

_____ 1. Plowing	a. The process of digging deep into the soil and turning it over.
_____ 2. Tilling	b. An agricultural method in which land is cleared and farmed for only a few years until the soil is depleted of nutrients.
_____ 3. Slash-and-burn agriculture (Shifting agriculture)	c. The preparation of soil through a variety of activities including plowing but also including stirring, digging, and cultivating.

MODULE 27: Irrigation and Pest Control Methods

Before You Read the Module

Focus on Learning Goals
Use the module learning goals to guide your reading. On a separate piece of paper, write down each goal and take notes to help you meet each learning goal. After studying this module, you should be able to:
- 27-1 describe the sources and locations of water used for irrigation.
- 27-2 explain the different types of irrigation.
- 27-3 describe how irrigation can cause waterlogging, salinization, and aquifer depletion.
- 27-4 identify the advantages and consequences of pest control methods.
- 27-5 describe pesticide resistance and the impacts of genetic engineering on crops.

Key Terms

Aquifer	Water footprint	Rodenticide
Unconfined aquifer	Furrow irrigation	Persistent pesticides
Confined aquifer	Flood irrigation	Nonpersistent pesticides
Water table	Spray irrigation	Integrated pest management (IPM)
Groundwater recharge	Drip irrigation	Pesticide resistance
Spring	Cone of depression	
Artesian well	Fungicide	

While You Read the Module
Answer the following questions as you read. Use a separate sheet of paper if necessary.

Module 27: Irrigation and Pest Control Methods

Groundwater and surface water are used for irrigation

1. Describe when irrigation is used for growing crops.

2. What is the largest use of water worldwide?

3. What is irrigation?

4. Define aquifer.

5. Define unconfined aquifer.

6. Define confined aquifer.

7. Define water table.

8. Figure 27.1: Explain when a confined aquifer can be recharged.

9. Define groundwater recharge.

10. Define spring.

11. Describe how humans obtained water from aquifers centuries ago.

12. Define artesian well.

13. Describe the age of water in both unconfined and confined aquifers.

Major Uses of Water

14. Define water footprint.

15. Figure 27.3: Compare water usage among nations.

16. Describe the amount of water used to produce grain and to produce beef.

Crop irrigation can be done by furrow, flood, spray, and drip irrigation

17. Describe furrow irrigation and the percent efficiency farmers can achieve using it.

18. Describe flood irrigation and the percent efficiency farmers can achieve using it.

19. Describe spray irrigation and the percent efficiency farmers can achieve using it.

20. Describe drip irrigation and the percent efficiency farmers can achieve using it.

There are three major adverse consequences of irrigation

Waterlogging

21. Using Figure 27.5a and 27.5b, what are the long-term impacts of flood irrigation?

Salinization

22. Define salinization.

Aquifer Depletion

23. Describe the significance of the Ogallala aquifer.

24. Identify and describe the negative impact to the Ogallala aquifer from 1950-2015.

25. Figure 27.7: List the states impacted by the Ogallala aquifer. Identify the state(s) with the greatest loss of water.

26. Define cone of depression.

27. Figure 27.8: Describe the negative impact of the cone of depression from diagram (a) to (b).

Pesticides reduce pests but have some adverse consequences

28. Describe the results of successful pest control. How is this achieved in developed countries?

29. Define fungicide.

30. Define rodenticide.

31. Define persistent pesticide and give an example.

32. Define nonpersistent pesticide.

33. Define integrated pest management.

Organisms can become pesticide resistant

34. Define pesticide resistance.

35. Figure 27.9: Describe the pesticide treadmill.

36. Describe the wide range of environmental effects of pesticides.

Potential Benefits of Genetic Engineering

37. Identify the insects that attack corn.

38. Describe *bacillus thuringiensis*, Bt.

39. Identify how Bt is used in corn.

40. Describe "Roundup Ready" gene and how it works.

41. Identify the benefits of Bt and HT corn.

Practice the Math: Calculating Per Capita Water Use

Read "Do the Math: Calculating Per Capita Water Use" on page 324. Try "Your Turn." For more math practice, do the following problem. Remember to show your work. Use a separate sheet of paper if necessary.

Germany and Turkey both have populations of 80 million people. The total annual water use in Germany is 6.59×10^{12} liters and in Turkey is 1.43×10^{13} liters. Calculate the per capita water use for each country and conclude which country has the larger water footprint.

Review Key Terms
Match the key terms on the left with the definitions on the right.

_____ 1. Aquifer

_____ 2. Unconfined aquifer

_____ 3. Confined aquifer

_____ 4. Water table

_____ 5. Groundwater recharge

_____ 6. Spring

_____ 7. Artesian well

_____ 8. Water footprint

_____ 9. Furrow irrigation

_____ 10. Flood irrigation

_____ 11. Spray irrigation

_____ 12. Drip irrigation

_____ 13. Cone of depression

_____ 14. Fungicide

_____ 15. Rodenticide

_____ 16. Persistent pesticides

_____ 17. Nonpersistent pesticides

_____ 18. Integrated pest management (IPM)

_____ 19. Pesticide resistance

a. A form of irrigation where an entire field is flooded with water.

b. An area surrounding a well that does not contain groundwater.

c. The process by which water from precipitation percolates through the soil into groundwater.

d. Total daily per capita use of fresh water for a country or the world.

e. Pesticides that remain in the environment for years to decades.

f. Pesticides that break down relatively rapidly, usually in weeks to months, and have fewer long-term effects but because they must be applied more often their overall environmental impact is not always lower than that of persistent pesticides.

g. A form of irrigation where the farmer digs trenches, or furrows, along the crop rows and fills them with water.

h. Porous rock covered by soil.

i. A pesticide that specifically targets rodents.

j. A form of irrigation where a slowly dripping hose on the ground or buried beneath the soil delivers water directly to the plant roots.

k. Pore spaces found within permeable layers of rock and sediment underneath the soil that store groundwater.

l. Water that naturally percolates up to the surface.

m. Surrounded by a layer of impermeable rock or clay, which impedes water flow to or from the aquifer.

n. An agricultural practice that uses a variety of techniques to minimize pesticide inputs.

o. A well created by drilling a hole into a confined aquifer.

p. The uppermost level at which the groundwater in a given area fully saturates the rock or soil.

q. A trait possessed by certain individuals that are exposed to a pesticide and survive.

r. A pesticide that specifically targets fungi (the plural of fungus).

s. A form of irrigation where water is pumped into an apparatus that contains a series of spray nozzles.

MODULE 28: Meat Production Methods and the Impacts of Overfishing

Before You Read the Module

Focus on Learning Goals
Use the module learning goals to guide your reading. On a separate piece of paper, write down each goal and take notes to help you meet each learning goal. After studying this module, you should be able to:

- 28-1 describe the benefits and consequences of different meat production methods.
- 28-2 describe the causes and consequences of overfishing.

Key Terms

Manure lagoon	Desertification	Concentrated animal feeding
Free range grazing	Fishery	operation (CAFO)
Nomadic grazing	Fishery collapse	
Overgrazing	Bycatch	

While You Read the Module
Answer the following questions as you read. Use a separate sheet of paper if necessary.

Module 28: Meat Production Methods and the Impacts of Overfishing

Concentrated animal feeding operations and free-range grazing each have their advantages and disadvantages

1. Identify the objectives of modern agribusiness in farming meat and poultry.

2. Describe why meat production is inherently less efficient than agricultural crop production.

High-Density Animal Farming

3. Define concentrated animal feeding operation (CAFO) and identify the animals that are raised this way.

4. Describe the benefits of CAFO farming.

5. Identify the adverse effects of CAFOs on the environment and the health of humans.

6. Explain the reason for concern about manure from concentrated animal feeding operations.

7. Define manure lagoon.

8. Identify the benefit of manure lagoons.

9. Identify detrimental effects of manure lagoons.

10. Describe the process of eating lower on the food chain.

Free-Range Grazing

11. Describe free range grazing.

12. List the advantages of raising free-range animals.

13. List the possible negative attributes of raising free-range animals.

14. Define nomadic grazing.

Overgrazing and Desertification

15. Define overgrazing

16. Define desertification.

17. Figure 28.5: Identify the regions that are most vulnerable to desertification.

Harvesting fish can result in fishery decline

18. Define fishery.

19. Describe how the tragedy of the commons applies to the world's ocean fish.

20. Explain this statement: "Fishers were working harder but catching fewer fish."

21. Define fishery collapse.

22. Identify the methods used to catch large numbers of fish.

23. Identify and describe the harmful effects of large-scale fishing.

24. Define bycatch.

More Sustainable Fishing

25. Figure 28.8: Describe the cause of the fishery collapse that occurred in the 1990s.

26. Describe what the United States Congress passed due to the fishery collapse.

27. Describe how Elinor Ostrom contributed to changing the Maine lobster fisheries.

28. Describe how consumers can identify the best sustainable fish to eat.

After You Read the Module

Review Key Terms
Match the key terms on the left with the definitions on the right.

_____	1.	Concentrated animal feeding operation (CAFO)	a. Allowing animals to graze outdoors on grass for most or all of their lifecycle.
_____	2.	Manure lagoon	b. Excessive grazing that can reduce or remove vegetation and erode and compact the soil.
_____	3.	Free range grazing	c. A large indoor or outdoor structure designed for maximum occupancy of animals and maximum output of meat.
_____	4.	Nomadic grazing	d. The decline of a fish population by 90 percent or more.
_____	5.	Overgrazing	e. Human-made pond lined with rubber built to handle large quantities of manure produced by livestock.
_____	6.	Desertification	f. Transformation of arable, productive, low-precipitation land to desert or unproductive land due to climate change or destructive land use such as overgrazing and logging.
_____	7.	Fishery	g. The unintentional catch of nontarget species while fishing.
_____	8.	Fishery collapse	h. A commercially harvestable population of fish within a particular ecological region.
_____	9.	Bycatch	i. The feeding of herds of animals by moving them to seasonally productive feeding grounds, often over long distances.

MODULE 29: Impacts of Mining

Before You Read the Module

Focus on Learning Goals
Use the module learning goals to guide your reading. On a separate piece of paper, write down each goal and take notes to help you meet each learning goal. After studying this module, you should be able to:
- 29-1 describe how natural resources are extracted from Earth through mining.
- 29-2 identify the ecological and economic impacts of mining.

Key Terms

Crustal abundance	Strip mining	Placer mining
Ore	Mine tailings	Subsurface mining
Metal	Open-pit mining	
Reserve	Mountaintop removal	

While You Read the Module
Answer the following questions as you read. Use a separate sheet of paper if necessary.

Module 29: Impacts of Mining

The distribution of mineral resources on Earth has social and environmental consequences

Abundance of Ores and Metals

1. Define crustal abundance.

2. Figure 29.1: Identify the top four elements in Earth's crust.

3. Define ore.

4. Define metals.

5. Describe the geologic process that forms ores.

6. Define reserve.

7. Describe a resource. Explain the difference between a resource and a reserve.

8. Propose a way in which humans could prevent metal reserves from running out as quickly as estimated.

Mining Techniques

Surface Mining

9. Define strip mining.

10. Define mine tailings.

11. Define open-pit mining.

12. Identify the world's largest open pit mine and its size.

13. Define mountaintop removal.

14. Define placer mining.

15. Give an example of placer miners.

Subsurface Mining

16. Define subsurface mining and describe how it is done.

Mining impacts our ecosystems

17. Table 29.2: Compare the effects of surface mining and subsurface mining.

18. Describe the environmental problems associated with construction of roads for mines.

19. Describe the environmental problems associated with mine tailings.

20. Describe the environmental problems associated with mountaintop removal.

21. Describe the environmental problems associated with placer mining.

22. Describe acid mine drainage.

Mining affects human safety and the economy

23. List the occupational hazards associated with subsurface mining.

24. Describe what happens when the demand for mineral resources continues to increase beyond the easily mined resources.

After You Read the Module

Review Key Terms
Match the key terms on the left with the definitions on the right.

_____	1. Crustal abundance	a.	The average concentration of an element in Earth's crust.
_____	2. Ore	b.	The removal of overlying vegetation and "strips" of soil and rock to expose underlying ore.
_____	3. Metal	c.	A mining technique in which the entire top of a mountain is removed with explosives.
_____	4. Reserve	d.	A mining technique that creates a large visible pit or hole in the ground.
_____	5. Strip mining	e.	Mining techniques used when the desired resource is more than 100 m (328 feet) below the surface of Earth.
_____	6. Mine tailings	f.	A concentrated accumulation of minerals from which economically valuable materials can be extracted.
_____	7. Open-pit mining	g.	Unwanted waste material created during mining including mineral and other rock residues that are left behind after the desired metals are removed from the ore.
_____	8. Mountaintop removal	h.	The process of looking for minerals, metals, and precious stones in river sediments.
_____	9. Placer mining	i.	An element with properties that allow it to conduct electricity and heat energy and to perform other important functions.
_____	10. Subsurface mining	j.	In resource management, the known quantity of a resource that can be economically recovered.

MODULE 30: Impacts of Urbanization and the Methods to Reduce Urban Runoff

Before You Read the Module

Focus on Learning Goals

Use the module learning goals to guide your reading. On a separate piece of paper, write down each goal and take notes to help you meet each learning goal. After studying this module, you should be able to:

- 30-1 explain urbanization and its effects on the environment.
- 30-2 describe the methods used to reduce urban runoff.

Key Terms

Urbanization	Saltwater intrusion	Sense of place
Urban area	Impervious surface	Urban runoff
Suburbs	Urban sprawl	
Exurbs	Urban blight	

While You Read the Module

Answer the following questions as you read. Use a separate sheet of paper if necessary.

Module 30: Impacts of Urbanization and the Methods to Reduce Urban Runoff

Increasing population density has many environmental implications

1. Describe how changing population density impacts the environment.

2. Define urbanization.

3. Define urban area.

4. What is the most populated city in the United States?

Developed and Developing Country Differences

5. Figure 30.1: Where will a majority of the world's population live by 2030?

6. Table 30.1: How many of the most populous cities are in developing countries?

7. List the environmental challenges of urban living for developed and developing countries.

Increasing Urbanization and Urban Sprawl in the United States

8. Define suburbs.

9. Define exurb.

10. Define saltwater intrusion.

11. Figure 30.3: Describe the cause of saltwater intrusion from (a) to (b).

12. Define impervious surface.

13. Describe two environmental problems with impervious surfaces.

14. How are impervious surfaces related to flooding?

Causes and Consequences of Urban Sprawl

15. Define urban sprawl.

16. Identify the visible characteristics of urban sprawl.

17. Describe the environmental impacts of urban sprawl.

18. Explain how modern transportation has facilitated the growth of suburbs.

19. Describe the four main causes of urban sprawl in the United States.

20. Define urban blight.

21. Figure 30.4: List the steps in the positive feedback loop of urban blight starting with the arrow indicating population shifts to suburbs.

Reducing Impacts of Urbanization and Urban Sprawl: Smart Growth

22. Describe how smart growth addresses urban sprawl.

23. Define sense of place.

24. List the basic principles of smart growth.

A variety of methods can reduce urban runoff

25. Define urban runoff.

26. Describe three methods for reducing urban runoff.

27. What happens to runoff in urban areas?

28. Describe what happens to urban runoff during heavy rain events.

29. Describe methods cities can use to reduce urban runoff pollutants.

After You Read the Module

Review Key Terms
Match the key terms on the left with the definitions on the right.

_____ 1. Urbanization

a. Similar to suburbs, but are not connected to any central city.

_____ 2. Urban area

b. An infiltration of salt water in an area where groundwater pressure has been reduced as a result of a cone of depression from extensive pumping of wells.

_____ 3. Suburbs

c. Areas that surround metropolitan centers.

_____ 4. Exurbs

d. Pavement or other surfaces that do not allow water penetration.

_____ 5. Saltwater intrusion

e. An area that contains more than 386 people per square kilometer (1,000 people per square mile).

_____ 6. Impervious surface

f. Runoff, water that does not evapotranspire or infiltrate the soil, that occurs in an urban area.

_____ 7. Urban sprawl

g. A lack of support for and deterioration of urban communities.

_____ 8. Urban blight

h. Urbanized areas that spread into rural areas.

_____ 9. Sense of place

i. The process of making an area more urban, which means increasing the density of people per unit area of land.

_____ 10. Urban runoff

j. The feeling that an area has a distinct and meaningful character.

MODULE 31: Ecological Footprints

Before You Read the Module

Focus on Learning Goals

Use the module learning goals to guide your reading. On a separate piece of paper, write down each goal and take notes to help you meet each learning goal. After studying this module, you should be able to:

- 31-1 explain the ecological footprint and what it tells us.
- 31-2 describe the carbon footprint and how it differs from the ecological footprint.

Key Terms

Ecological footprint   Carbon footprint

While You Read the Module

Answer the following questions as you read. Use a separate sheet of paper if necessary.

Module 31: Ecological Footprints

The ecological footprint is an environmental assessment tool

Ecological Footprint: Origins and Purpose

1. Define ecological footprint.

2. Describe why calculating an ecological footprint may not be simple.

Calculating the Ecological Footprint

3. How are total ecological footprints calculated?

4. How is an ecological footprint similar to the IPAT formula?

5. Describe the differences in resource use for individuals living in developed versus developing countries.

6. What does it mean if an ecological footprint is larger than 1.6 ha?

The carbon footprint calculates carbon dioxide emitted into the atmosphere

7. Define carbon footprint.

8. What two aspects of carbon footprint go into the calculation?

9. What two categories account for up to 60% of a person's carbon footprint?

10. Figure 31.3: Describe one way that people can lower their carbon footprint.

After You Read the Module

Review Key Terms
Match the key terms on the left with the definitions on the right.

_____	1. Ecological footprint	a. A measure of the total carbon dioxide and other greenhouse gases emissions from the activities, both direct and indirect, of a person, country, or other entity.
_____	2. Carbon footprint	b. A measure of the area of land and water an individual, population, or activity requires to produce all the resources it consumes and to process the waste it generates.

MODULE 32: Introduction to Sustainability

Before You Read the Module

Focus on Learning Goals

Use the module learning goals to guide your reading. On a separate piece of paper, write down each goal and take notes to help you meet each learning goal. After studying this module, you should be able to:

- 32-1 explain sustainability.
- 32-2 describe some of the environmental indicators of sustainability.

Key Terms

Sustainability
Sustainable development

Maximum sustainable yield
(MSY)

Environmental indicators
Anthropogenic

While You Read the Module

Answer the following questions as you read. Use a separate sheet of paper if necessary.

Module 32: Introduction to Sustainability

Sustainability is the most comprehensive assessment of environmental impact

1. Define sustainability.

The Impact of Consumption on the Environment

2. What effect does increased consumption have on the environment?

3. Figure 32.1: Compare the population from 1850-1950 to 1950-2011.

4. What does it mean to live sustainably?

5. Define sustainable development.

6. Figure 32.2: Explain why biking to work or school considered a sustainable practice.

Maximum Sustainable Yield

7. Define maximum sustainable yield.

8. Figure 32.3: Identify the type of growth curve shown.

9. Describe the effect of unlimited deer hunting.

10. Figure 32.3: Describe the point on the graph where population rates should increase the fastest.

11. Describe the result if maximum sustainable yield is surpassed in the forest.

12. Describe a problem with using maximum sustainable yield in environmental policy.

Scientists use environmental indicators to monitor the state of Earth

13. What question do environmental scientists investigate when examining sustainability?

14. Define environmental indicators.

15. List the five global-scale environmental indicators that this textbook will focus on.

Biodiversity

16. What does a decrease in overall biodiversity tell us?

Food Production

17. Describe the need for food production.

18. Describe how science and technology have been used to increase the amount of food that is produced on a given area of land.

19. Figure 32.4: How have grain production, utilization, and reserves changed from 2011 to 2021?

20. List the factors that influence grain production.

Average Global Surface Temperature and Carbon Dioxide Concentrations

21. How long has Earth's temperature been relatively constant?

22. What keeps Earth's temperature so constant?

23. Figure 32.5: Notice the solar energy entering the heat-trapping greenhouse gases. Identify and describe how the heat moves in the diagram.

24. What gas contributes most to the warming of the atmosphere?

25. Figure 32.6: Describe the correlation of the carbon dioxide levels with the global temperatures shown on the graph.

26. Define anthropogenic.

27. What do scientists believe is the cause in the rise of atmospheric carbon dioxide?

Human Population

28. What is the approximate current human population?

29. What is Earth's projected human population?

30. Can the planet sustain so many people?

Resource Depletion

31. Identify some of Earth's natural resources that are finite and cannot be renewed or reused.

32. Identify some of Earth's natural resources that are finite but can be reused or recycled.

33. Identify an example of Earth's renewable resources. Can you think of another example?

34. Explain how development influences lifestyle.

35. Figure 32.8: Consider the pie charts. Which shows the greatest difference in developed versus developing nations? Which other chart demonstrates a large imbalance of resource use between developed and developing countries?

36. Table 32.2: Identify and explain what key global indicators you are most concerned about after reading the module.

Read "Do the Math: Rates of Forest Clearing" on page 367. Try "Your Turn." For more math practice, do the following problem. Remember to show your work. Use a separate sheet of paper if necessary. Environmental organizations have yielded a range of estimates of the amount of forest clearing that is occurring in the Brazilian Amazon. Convert the first two estimates into hectares per day and compare the three estimates:

- Estimate 1: 15 acres per minute

- Estimate 2: 22,000 acres per day

- Estimate 3: 8,000 ha per day

After You Read the Module

Review Key Terms
Match the key terms on the left with the definitions on the right.

_____ 1. Sustainability	a. Development that balances current human well-being and economic advancement with resource management for the benefit of future generations.
_____ 2. Sustainable development	b. Describe the current state of an environmental system or the Earth.
_____ 3. Maximum sustainable yield (MSY)	c. Derived from human activities.
_____ 4. Environmental indicators	d. Being able to use a resource or engage in an activity now without jeopardizing the ability of future generations to engage in similar activities later.
_____ 5. Anthropogenic	e. The largest quantity of a renewable resource that can be harvested indefinitely.

MODULE 33: Integrated Pest Management and Sustainable Agriculture

Before You Read the Module

Focus on Learning Goals
Use the module learning goals to guide your reading. On a separate piece of paper, write down each goal and take notes to help you meet each learning goal. After studying this module, you should be able to:
- 33-1 describe the goals and techniques of integrated pest management (IPM).
- 33-2 explain the objectives of sustainable agriculture practices.

Key Terms

Crop rotation	Agroforestry	No-till agriculture
Intercropping	Windbreaks	Green manure
Biocontrol	Strip cropping	Limestone
Natural predators	Contour plowing	Rotational grazing
Sustainable agriculture	Terracing	Organic agriculture
Soil conservation	Perennial plants	Delaney Clause

While You Read the Module
Answer the following questions as you read. Use a separate sheet of paper if necessary.

Module 33: Integrated Pest Management and Sustainable Agriculture

Integrated pest management increases agricultural output while minimizing environmental harm

Crop Rotation and Intercropping

1. Define crop rotation and give an example.

2. Define intercropping and give an example.

Biocontrol and Natural Predators

3. Define biocontrol.

4. Define natural predators.

5. Describe how IPM practitioners still utilize pesticides.

6. Figure 33.3: Describe how IPM training in Indonesia assisted in growing crops.

7. Describe the tradeoffs of IPM.

Sustainable agriculture reduces environmental impacts

8. Define sustainable agriculture.

9. Define soil conservation.

Traditional Sustainable Farming Techniques

10. What practices are used in sustainable agriculture?

11. Figure 33.4: Explain whether the farming method in each photograph would increase or encourage higher levels of biodiversity in the environment.

12. Define agroforestry.

13. Define windbreaks. How are they beneficial?

14. Describe the advantages of agroforestry.

15. Define strip cropping. Provide an example.

16. Define contour plowing.

17. Describe utilizing a cover crop.

18. Define terracing. How does it prevent erosion?

Modern Sustainable Farming Techniques

19. Define perennial plants.

20. What are the advantages of perennial plants?

21. Identify the species researchers are trying to convert to perennial crops.

22. Define no-till agriculture.

23. Describe the advantages of no-till agriculture.

24. Describe a disadvantage of no-till agriculture.

Soil Additives and Rotational Grazing

25. How do sustainable farmers add nutrients to the soil?

26. Define green manure and provide advantages of its use.

27. What are the disadvantages of green manure?

28. What happens to base cations when crops are removed from agricultural fields?

29. Define limestone and describe why a farmer might add it to their fields.

30. Define rotational grazing.

Organic Agriculture

31. Define organic agriculture.

32. List the basic principles of organic agriculture.

33. Define the Delaney Clause.

34. Explain why organic farms tend to be small.

35. Identify and describe the adverse environmental consequences of weed-free carrots.

Practice the Math: Comparing Carbon Accumulation in Tilled and No-Till Agricultural Soils

Read "Do the Math: Comparing Carbon Accumulation in Tilled and No-Till Agricultural Soils" on page 375. Try "Your Turn." For more math practice, do the following problems. Remember to show your work. Use a separate sheet of paper if necessary.

A group of scientists also compared farming practices in soils in Nebraska, Michigan, Ohio, and Kentucky. They recorded the amount of total carbon in the 0- to 20-cm layer of soil using grams of carbon per square meter.

	Total C (g C m-2) in the 0- to 20-cm soil layer			
Treatment	Nebraska	Michigan	Ohio	Kentucky
Native Grassland	4,090	2,944	4,008	5,036
Tilled Farmland	2,907	2,209	3,380	3,125
No-Till Farmland	3,428	2,444	3,806	3,742

Data from: J. Six et al. *Soil Sci. Soc. Am. J.* 63:1350–1358. 1999.

"Do the Math," concluded that tilled farmland contains 71.08 percent of the carbon content of native grassland and no-till farmland contains 83.81 percent.

(a) Determine the amount of carbon in tilled and no-till farmland as a percentage of the total carbon content in native grasslands in Michigan, Ohio, and Kentucky.

(b) For each state determine if tilled farming or no-till was more effective at maintaining the carbon content of soil.

(c) Which state and farming practice yielded the highest carbon maintenance efficiencies? Which state and farming practice yielded the lowest carbon maintenance efficiencies?

Review Key Terms

Match the key terms on the left with the definitions on the right.

_____ 1. Crop rotation

_____ 2. Intercropping

_____ 3. Biocontrol

_____ 4. Natural predators

_____ 5. Sustainable agriculture

_____ 6. Soil conservation

_____ 7. Agroforestry

_____ 8. Windbreaks

_____ 9. Strip cropping

_____ 10. Contour plowing

_____ 11. Terracing

_____ 12. Perennial plants

_____ 13. No-till agriculture

_____ 14. Green manure

_____ 15. Limestone

_____ 16. Rotational grazing

_____ 17. Organic agriculture

_____ 18. Delaney Clause

a. An agricultural method used in fields of annual crops where farmers do not till or plow the soil between seasons.

b. Plant material deliberately grown in a field with the intention of plowing it under at the end of the season.

c. A shortened term for biological control, it uses biological organisms to control agricultural pests.

d. A calcium carbonate sedimentary rock that has been ground up or crushed for easy application as a fertilizer.

e. Predators that occur naturally in the environment.

f. The rotation of farm animals to different pastures and fields to prevent overgrazing.

g. Fulfills the need for food and fiber while enhancing the quality of the soil, minimizing the use of nonrenewable resources, and allowing economic viability for the farmer.

h. The prevention of soil erosion while simultaneously increasing soil depth and increasing the nutrient content and organic matter content of the soil.

i. A clause in the Food, Drug, and Cosmetic Act designed to prevent potentially harmful cancer-causing food ingredients.

j. Plowing and harvesting parallel to the topographic contours of the land.

k. An agricultural method of planting crops with different spacing and rooting characteristics in alternating sets of rows to prevent soil erosion.

l. An agricultural technique in which trees and vegetables are intercropped.

m. An agricultural technique that literally plants tall objects that "break" the wind and prevent soil erosion.

n. A crop-planting strategy in which different types of crop species are planted from season to season or year to year on the same plot of land.

o. An agricultural technique where farms shape sloping land into step-like terraces that are flat.

p. An agricultural technique that calls for physical spacing of different crops growing at the same time, in close proximity to one another, to promote biological interaction.

q. Plants that live for multiple years and do not need to be replanted at the beginning of each growing season.

r. The production of crops in a way that sustains or improves the soil, without the use of synthetic pesticides or fertilizers.

MODULE 34: Aquaculture and Sustainable Forestry

Before You Read the Module

Focus on Learning Goals
Use the module learning goals to guide your reading. On a separate piece of paper, write down each goal and take notes to help you meet each learning goal. After studying this module, you should be able to:
- 34-1 explain the environmental benefits and consequences of aquaculture.
- 34-2 describe the methods for reducing human impacts on forest tree removal.

Key Terms

Reforestation   Sustainable forestry   Prescribed burn

While You Read the Module
Answer the following questions as you read. Use a separate sheet of paper if necessary.

Module 34: Aquaculture and Sustainable Forestry

Aquaculture is increasing in importance and its environmental impacts are less

1. Describe why there is a great need to find alternative methods for obtaining fish for food consumption.

2. Describe the requirements of producing seafood through aquaculture.

Aquaculture Systems: Nets, Ponds, Recirculating Aboveground Tanks

3. Figure 34.2: Describe possible environmental concerns of aquaculture shown in the photograph.

4. Describe the similarities in various aquaculture operations.

5. Describe the advantages of aquaculture.

6. Describe the environmental concerns with aquaculture.

Sustainable forestry uses principles found in agriculture to minimize the effects of logging

7. Define reforestation.

8. Describe the environmental benefits of reforestation.

9. Define sustainable forestry.

10. What are the goals when removing trees in sustainable forestry?

Other Sustainable Logging Techniques

11. What are the most important first steps in sustainable forestry?

12. How might foresters attempt to exclude the introduction of insect pests to forests?

13. What are pest control methods if a forest pest is already present?

Fire Management

14. Identify the importance of ecosystem fires.

15. Describe the negative effect of fire suppression.

16. Define prescribed burn.

17. List some of the causes of fires on public or private lands.

18. Describe what researchers and forest managers observed about the Yellowstone National Park fire of 1988.

Visual Representation 5: Land use: problems and solutions

19. Describe how agriculture causes environmental problems in warm climate locations.

20. Describe solutions to the problems you listed above.

21. How can fishing be an example of tragedy of the commons? What is an appropriate solution?

22. How can clear-cutting be an example of tragedy of the commons? What is an appropriate solution?

23. Describe how IPM and organic agriculture can be used as solutions to pests. Why are pesticides problematic?

24. Describe why urban runoff is problematic and possible solutions.

Pursuing Environmental Solutions

Urban Agriculture

25. List the ecosystem services of an urban farm.

26. Describe the urban benefits of farming in the city.

27. Identify what the USDA defines as urban agriculture.

28. Identify the circumstances in Detroit that led to a need for urban agriculture.

29. Explain why the D-Town Farm was established.

30. Describe how the urban gardening grew over the last two decades.

31. Identify where the urban garden food is used.

32. Explain the statement "Eating is an agricultural act."

33. List possible economic benefits that urban gardening can provide to individuals and the community.

Science Applied 5

How Do We Define Organic Food?

34. Describe the reasons some people prefer to buy organic food.

How did the organic food movement begin?

35. Identify what organic food enthusiasts desire.

36. Figure SA5.2: Use the graph to describe the trend in land devoted to organic farming.

37. Identify the states with the most land devoted to organic farmland.

38. What percentage of food sales is organic?

What does it mean to be organic?

39. Describe what the USDA determined to be organic agriculture.

40. Identify the methods organic farmers can use to control insects.

41. Explain the results of failed inspections of food producers and processors.

42. Explain how certification of organic foods is conducted.

Does organic food mean family farms?

43. Identify the grocery retailers that are helping drive the organic food industry and the significance of their involvement.

44. What changes has the USDA made in the number of non-organic and synthetic additives permitted in foods that are labeled organic?

What are the four different USDA organic designations?

45. What are the four different organic designations?

What does the organic label not mean?

46. List three things an organic label does NOT indicate.

47. Describe the results of researchers when they studied organic versus traditional food.

After You Read the Module

Review Key Terms
Match the key terms on the left with the definitions on the right.

_____	1. Reforestation	a. A methodology for managing forests so they provide wood while also providing clean water, maximum biodiversity, and maximum carbon sequestration in both trees and soil.
_____	2. Sustainable forestry	b. The natural or intentional restocking of trees after clear-cutting to repopulate the forest, reduce erosion, and begin the process of removing carbon dioxide from the atmosphere.
_____	3. Prescribed burn	c. When a fire is deliberately set under controlled conditions, thereby decreasing the accumulation of dead biomass on the forest floor.

UNIT 5 Review Exercises

Check Your Understanding

Review "Learning Goals Revisited" at the end of each module in Unit 5 of your textbook. Compare the notes you took while reading each module. Complete these exercises to review the unit. Use a separate sheet of paper if necessary.

1. Explain the advantages and disadvantages of both clear cutting and selective cutting.

2. How can irrigation contribute to soil degradation?

3. List advantages and disadvantages of using synthetic fertilizers.

4. Describe the advantages and disadvantages of plowing or tilling.

5. Describe one advantage and one disadvantage of each type of irrigation.

6. Compare and contrast CAFOs and free-range grazing for animal production.

7. Describe one impact subsurface mining can have on human health.

8. Compare and contrast carbon, water, and ecological footprints.

9. Describe how biocontrol and natural predators can be a component of an IPM.

10. Describe aquaculture.

Practice for Free-Response Questions

Complete this exercise to build and practice the skills you will need to answer free-response questions on the exam. Use a separate sheet of paper if necessary.

1. Cod fisheries have long been a battlefield between conservationists and fisherman who depend on the cod for their livelihood. Between 1850 and 1950 cod populations in the North Atlantic fluctuated between 100,000 tons and 300,000 tons. Following tough environmental regulations, the populations peaked at 800,000 tons in the 1970s. Due to the population growth, regulations were lifted, and the population plummeted to under 10,000 in 2000.

    (a) Explain the environmental concept that describes overuse of a common resource.

    (b) Calculate the percent decline in cod fisheries from 1850 to 2000. Assume there were 100,000 tons of cod in 1850.

2. Explain two unintended environmental impacts of industrialized agriculture.

3. An AP® Environmental Science student states that there are no benefits to synthetic fertilizers. Provide one piece of evidence that might refute their claim.

4. Explain how over time, the effectiveness of a pesticide decreases and ultimately results in the pesticide treadmill.

5. All forms of mining affect the environment.

    (a) Describe ONE impact mining has on air pollution.

    (b) Describe ONE impact mining has on water pollution.

    (c) Describe ONE impact mining has on biodiversity.

6. Many large urban areas in the United States lack zoning restrictions which results in large numbers of parking lots, paved roads, and other impervious surfaces.

   (a) Describe one environmental problem associated with impervious surfaces.

   (b) Provide one solution for the problem you listed above and justify its use.

7. Explain how agroforestry can be both a pest control method and be used for soil conservation.

8. Explain how aquaculture and sustainable forestry may be considered possible solutions to tragedy of the commons problems such as fishing in the oceans or harvesting trees in a forest.

Unit 5 Multiple-Choice Review Exam

1. Which would be an example of the tragedy of the commons?
   (a) air pollution caused by a factory
   (b) deforestation on private land
   (c) a school converting its football field into a parking lot
   (d) a farmer draining a lake on his land

2. Which is an example of a positive externality?
   (a) post-holiday sale prices
   (b) feeling relaxed after a vacation
   (c) pollution from automobile exhaust
   (d) an existing mangrove forest preventing coastal storm damage

3. Which of the following represents an activity that is permitted on Bureau of Land Management, United States Forest Service, National Park Service, and Fish and Wildlife Service lands?
   (a) Mining
   (b) Recreation
   (c) Timber Harvesting
   (d) Hunting

4. Which of the following is a consideration for managing rangelands?
   (a) Rangelands are primarily used for raising tree plantations for paper production.
   (b) Rangelands may be arid and prone to erosion due to overgrazing.
   (c) Rangelands can support an infinite number of organisms.
   (d) Rangelands are typically managed for recreation of tourists.

5. A logging company selectively clears a job site and removes only enough large trees to let light reach the younger trees. What is the purpose of this practice?
   (a) to prevent any negative externalities
   (b) to create a positive externality, despite logging trees in the forest
   (c) to avoid a tragedy of the commons
   (d) to harvest using the principle of maximum sustainable yield

6. The majority of land in the United States is used for
   (a) timber production.
   (b) grassland and grazing land.
   (c) forests and grazing land.
   (d) recreational and wildlife land.

7. Which is the most profitable way to harvest trees?
   (a) selective cutting
   (b) clear-cutting
   (c) waterlogging
   (d) using horses instead of machines

8. Prescribed burns are used to
   (a) destroy invasive species.
   (b) reduce herbicide use.
   (c) clear land economically.
   (d) reduce the accumulation of dead biomass.

9. Which is a concern about tree plantations?
   (a) cost of harvesting trees
   (b) lack of biodiversity
   (c) introduction of invasive species
   (d) difficulty with fire control

10. Which is an example of urban sprawl?
   (a) a heavily populated downtown business center
   (b) rangeland for grazing animals
   (c) suburbs with expanding neighborhoods
   (d) a large shopping complex in a city railway station

11. Which is a consequence of urban sprawl?
   (a) large homes close together
   (b) housing and retail shops separated by miles of road
   (c) decreased traffic congestion
   (d) decreased gasoline use

12. Which of the following is an advantage of smart growth?
   (a) Decreased habitat fragmentation due to less outward expansion.
   (b) Increased reliance on automobiles for transportation
   (c) Increased amounts of surface water runoff due to paved roads.
   (d) Lack of ability to use public transportation due to outward expansion.

13. Which of the following is an unintended consequence of aquaculture?
   (a) It provides a supply of protein for people worldwide.
   (b) It alleviates some of the pressure on over-exploited fisheries.
   (c) It boosts the economies of many developing countries.
   (d) It can lead to spread of disease and waste into wild fish populations.

14. Which of the following is an advantage of the Green Revolution?
   (a) Decreased crop yields due to declining soil fertility.
   (b) Increase use of fertilizers and pesticides that can run off into nearby water sources
   (c) Increase use of mechanization that uses fossil fuels.
   (d) Increase in crop yields for a growing human population.

15. Which typifies a pesticide treadmill?
   (a) A farmer uses biological pest controls and the biological agent reproduces.
   (b) A homeowner uses excessive fertilizer.
   (c) Pests reproduce exponentially.
   (d) A new pesticide must be used because of pest resistance.

16. Desertification is happening most rapidly in which location?
   (a) North America
   (b) Africa
   (c) Europe
   (d) Asia

17. Which of the following is a disadvantage of utilizing an Integrated Pest Management System?
    (a) Pesticide runoff can contaminate nearby water sources.
    (b) Farmers must constantly monitor their crops, leading to increased yields.
    (c) Farmers must be trained in using IPM and invest heavy amounts of time in the process.
    (d) Farmers and workers are less exposed to pesticides.

Use the following figure to answer question 18.

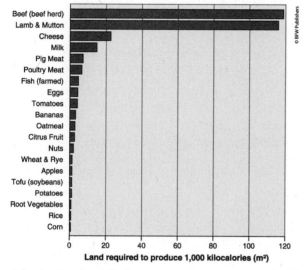

18. Approximately how many meters squared of land does it require to raise 1,000 kilocalories of beef?
    (a) 60 m²
    (b) 80 m²
    (c) 118 m²
    (d) 120 m²

19. CAFOs enable large quantities of meat to be raised on smaller plots of land. Which of the following is a characteristic of CAFOs?
    (a) increased strains of antibiotic-resistant microorganisms
    (b) minimal production of wastes that are easily contained
    (c) free-range chicken and beef
    (d) smaller quantities of animals that are frequently moved to new plots

20. What is one way to alleviate some of the pressure on overexploited fisheries?
    (a) Increase aquaculture practices.
    (b) Use long line fishing practices.
    (c) Encourage purse-seine fishing.
    (d) Use dragnets.

21. Confined aquifers are
    (a) polluted more easily than unconfined aquifers.
    (b) covered by impermeable rock.
    (c) accessed only through pumping.
    (d) covered by porous rock and soil.

22. The Ogallala aquifer is
    (a) the largest aquifer in the United States.
    (b) the largest aquifer in the world.
    (c) heavily polluted.
    (d) contaminated by saltwater intrusion.

23. Which type of irrigation is the most efficient?
    (a) spray irrigation
    (b) flood irrigation
    (c) furrow irrigation
    (d) drip irrigation

24. Irrigation has enabled increased crop production in areas that may not have sufficient rainfall for agriculture. Which of the following is an unintended consequence of irrigation?
    (a) underdrawing aquifers resulting in excessive recharging
    (b) pesticide use that can contaminate aquifers and other groundwater
    (c) salinization caused by salts left behind
    (d) overuse of aquifers that results in freshwater spilling out, contaminating the oceans

Use the following figure to answer questions 25 and 26.

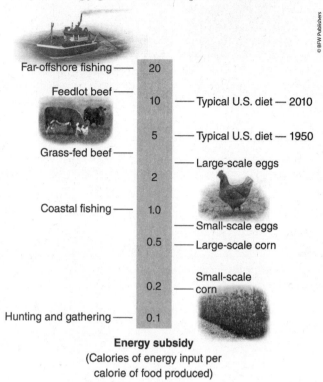

Far-offshore fishing —— 20

Feedlot beef —— 10 —— Typical U.S. diet — 2010

5 —— Typical U.S. diet — 1950

Grass-fed beef —— —— Large-scale eggs

2

Coastal fishing —— 1.0

—— Small-scale eggs

0.5 —— Large-scale corn

0.2 —— Small-scale corn

Hunting and gathering —— 0.1

Energy subsidy
(Calories of energy input per
calorie of food produced)

25. Which method of food production requires the greatest energy subsidy?
  (a) hunting and gathering
  (b) small-scale eggs
  (c) far-offshore fishing
  (d) feedlot beef

26. Which statement is best supported by the data?
  (a) Grass-fed beef has a smaller subsidy than feedlot beef.
  (b) Diets high in fish will always have lower subsidies than diets high in beef.
  (c) Large scale food production never requires as much energy as small-scale production.
  (d) Small scale food production typically requires the least amount of energy input.

TABLE 29.1	Approximate supplies of metal reserves remaining, assuming the same rate of metal use and same rate of recycling	
Metal	**Global reserves remaining (years)**	**U.S. reserves remaining (years)**
Aluminum (Al)	100	2
Copper (Cu)	45	20
Lead (Pb)	20	20
Zinc (Zn)	20	15
Gold (Au)	15	15
Nickel (Ni)	35	5
Cobalt (Co)	50	95
Manganese (Mn)	70	0
Chromium (Cr)	15	0

(Data from U.S. Geological Survey Mineral Commodity Summaries 2021, https://pubs.usgs.gov/periodicals/mcs2021/mcs2021 .pdf; Bauxite and Alumina Report (USGS), https://pubs.usgs.gov/periodicals/mcs2021/mcs2021-bauxite-alumina.pdf)

27. Based on global reserves, how many years do we have left to mine gold at current rate of use and current recycling levels?
    (a) 15 years
    (b) 35 years
    (c) 50 years
    (d) 100 years

28. Technology demands require the use of natural resources that must be mined. One negative impact of mining on water quality is that
    (a) particulate matter reacts with water to decrease turbidity of water sources.
    (b) road construction leads to habitat fragmentation.
    (c) fossil fuels can react with water sources to cause acidification.
    (d) tailings may percolate and contaminate nearby water sources.

29. In 2020, approximately 55% of the world's population lived in urban areas. In more developed countries, 75% of the population lived in urban areas, compared with only 49% of those individuals living in less developed countries. It is estimated that in 2030, 60% of the world's population will live in urban areas, with 81% of those in more developed countries living in urban areas, and 55% of those in less developed countries now occupying urban areas. What will be the percent increase in people in less developed countries living in urban areas in 2030 as compared to the number of people in 2020?
    (a) 7.4%
    (b) 8.00%
    (c) 9.00%
    (d) 12.25%

30. Which of the following gives a correct description of a possible solution to urban runoff?
    (a) Impervious pavement allows for high volumes of water to runoff.
    (b) Rooftop gardens absorb water that would otherwise runoff.
    (c) Taller buildings with more occupants mean more roads and buildings should be constructed.
    (d) Removal of trees and city parks increases the amount of soil and water runoff.

UNIT 6

Energy Resources and Consumption

Unit Summary

In this unit you will further your understanding of energy transformations (Big Idea 1), energy sources, energy efficiency, conservation, and non-renewable and renewable energy sources. You will learn how to analyze the pros and cons of each energy source and understand that energy supply for a nation requires a multi-faceted approach. This unit contains many opportunities for math practice and by the end of the unit you should be comfortable with energy math calculations.

MODULES IN THIS UNIT

Module 35: Renewable and Nonrenewable Resources and Global Energy Consumption
Module 36: Fuel Types and Uses
Module 37: Distribution of Natural Energy Resources: Fossil Fuels
Module 38: Nuclear Power
Module 39: Biomass and Solar Energies and Hydroelectric Power
Module 40: Geothermal Energy and Hydrogen Fuel Cells
Module 41: Wind Energy and Energy Conservation

Unit Opening Case: *All Energy Use Has Consequences*

This case study discusses some of the risks associated with use of nonrenewable fossil fuels. It describes the California coast pipeline accident of 2021, the *Exxon Valdez* spill in 1989, the BP Deepwater Horizon oil rig blowout in 2010, and the events at the Fukushima nuclear power plant in Japan. The United States consumes more energy (EJ) per capita than any other country in the world. The technological progress associated with increased energy consumption provides many benefits, but our dependence on fossil fuels for energy is not without long-term costs.

Do the Math

This unit contains the following "Do the Math" boxes to help prepare you for calculation questions you might encounter on the exam.
- "Comparing Energy Content in Fuels" (page 402)
- "Efficiency of Travel" (page 417)
- "Calculating Energy Supply" (page 422)
- "Comparing the Efficiencies of Two Systems" (page 432)
- "Calculating Half-Lives" (page 439)
- "Calculating Home PV Generation and Capacity Factor" (page 451)
- "Energy Star" (page 472)

To make sure you understand the concepts and techniques presented in these boxes, do the practice problems presented in the text as well as the additional "Practice the Math" problems that appear in Module 35, Module 36, Module 37, Module 38, Module 39, and Module 41 of this study guide.

MODULE 35: Renewable and Nonrenewable Resources and Global Energy Consumption

Before You Read the Module

Focus on Learning Goals

Use the module learning goals to guide your reading. On a separate piece of paper, write down each goal and take notes to help you meet each learning goal. After studying this module, you should be able to:
- 35-1 describe nonrenewable energy resource characteristics.
- 35-2 describe renewable energy resource characteristics.
- 35-3 describe trends of energy use worldwide and in the United States.
- 35-4 explain the importance of energy efficiency and conservation.

Key Terms

Fossil fuels
Nonrenewable energy resource
Renewable energy resources
Potentially renewable
Nondepletable

Commercial energy sources
Subsistence energy sources
Energy intensity
Fossil fuel combustion
Hubbert curve
Peak oil

Energy conservation
Energy efficiency
Energy return on energy investment (EROEI)

While You Read the Module

Answer the following questions as you read. Use a separate sheet of paper if necessary.

Unit Opening Case: All Energy Use Has Consequences

1. Describe the age of the carbon that is released when fossil fuels are burned.

2. List the negative effects from the oil spilled off the California coast in October of 2021.

3. List the negative consequences of the oil spill from the British Petroleum Deepwater Horizon oil extraction platform in the Gulf of Mexico in 2010.

4. List the negative consequences from the Exxon Valdez oil spill in 1989.

5. Identify other ways that oil spills can occur.

6. Describe ways in which coal mining is risky to workers.

7. Identify the consequences of the production of natural gas.

8. Describe two nuclear disasters.

9. Identify how much energy is used, on average, per person in the United States.

Module 35: Renewable and Nonrenewable Resources and Global Energy Consumption

Nonrenewable energy resources are finite and important

10. Define fossil fuels.

11. List the three types of fossil fuels.

12. Identify the environment where anaerobic activity takes place for fossil fuels to form. Describe why decomposers can't break down all the detritus.

13. Describe how the detritus is chemically changed into combustible materials.

14. Explain why combusting fossil fuels could have negative effects on the atmosphere and climate.

15. Define nonrenewable energy resources.

16. Identify the value of one gigajoule (GJ).

17. Identify the value of one exajoule (EJ).

18. Identify the value of a quad.

19. Identify the best candidates for replacing nonrenewable fossil fuels.

Renewable energy resources are infinite and becoming more important

20. Define renewable energy resources.

21. Define potentially renewable.

22. Define nondepletable.

23. Figure 35.1: List potentially renewable resources.

24. Figure 35.1: List nondepletable resources.

25. Identify and explain the energy resources currently used in some developing parts of the world.

Trends in energy use are changing around the world and in the United States

Worldwide Patterns of Energy Use

26. Figure 35.2: List the worldwide energy sources from greatest consumption to least and include the percentages.

27. Figure 35.3: Compare and contrast total annual energy consumption and per capita annual energy consumption in China and the United States.

28. Figure 35.3: Describe why some countries have a high total energy consumption but a much lower per capita energy consumption.

29. Define commercial energy sources.

30. Define subsistence energy source.

31. Describe what causes the change in energy demand in different countries.

Patterns of Energy Use in the United States

32. Describe how energy use changed from about 1875 to the early 1900s.

33. Figure 35.4: Explain the changes in the importance of coal and natural gas in recent years in the United States.

34. Figure 35.4: Explain the cause of the sharp decline of energy use in 2020.

35. Identify the energy inputs of the U.S. energy system.

36. Identify the energy outputs of the U.S. energy system.

37. Figure 35.5: Identify the percentage of United States energy consumption that is fossil fuels.

38. Identify the percentage of energy the United States produces for itself.

39. List the generating sources used for regional energy in the United States.

40. Identify and describe seasonal differences in energy usage.

Quantities of Fossil Fuels in the United States and Worldwide

41. Figure 35.6: Describe where the petroleum can be found.

42. List the organizations that assess the estimated reserves of fuel and project how much of each fuel is recoverable.

43. Table 35.2: List the fuel resources in the table by most to least years remaining for the United States and the world.

44. Figure 35.7: Describe the trend of energy use per capita in the United States since 1990.

45. Define energy intensity.

46. Define fossil fuel combustion.

47. Explain why our use of energy is not lower than it once was even though our use of energy has become more efficient.

48. Define the Hubbert curve.

49. Figure 35.8: Explain the reason Hubbert used two lines on the Hubbert curve.

50. Define peak oil.

51. Describe Hubbert's prediction from 1969.

52. Explain why some environmental scientists don't think it is necessary to determine exactly how much oil is left.

The Future of Fossil Fuel Use

53. Identify the changing usage patterns of energies.

We can use less energy through conservation and increased efficiency

54. Define energy conservation.

55. Define energy efficiency.

56. Explain why energy conservation and efficiency are the better options for maximizing the use of energy resources.

Different Forms of Energy

57. Describe the best fuel for transportation and explain why.

Quantifying Energy Efficiency

58. Describe energy efficiency.

59. Figure 35.9: Identify the purpose and significance of the red arrows in the diagram.

60. Identify how much energy is lost as heat or undesired outputs during coal-burning electricity generation.

61. Describe some of the other ways energy is lost in the coal generating electricity process.

62. Define energy return on energy investment (EROEI). Identify the equation.

63. Is EROEI the most efficient at higher or lower values?

Practice the Math: Comparing Energy Content in Fuels

Read "Do the Math: Comparing Energy Content in Fuels" on page 402. Try "Your Turn." For more math practice, do the following problem. Remember to show your work. Use a separate sheet of paper if necessary.

A therm is a unit used to measure the amount of natural gas a system uses and is equivalent to 100 cubic feet of natural gas. A British thermal unit (Btu) is the amount of energy needed to raise the temperature of one pound of water one degree in Fahrenheit.

1 therm = 100,000 Btu. 1 Btu = 1,055 joules. 1kWh = 3,600,000 J

Homes in the southern areas of the United States often use electricity for air conditioning and running most appliances, and they use natural gas for their furnace, hot water heater, and stovetop. An average home in the South uses 90 therms of natural gas and 13,000 kWh of electricity per year. Does this home use more energy in natural gas or in electricity in one year?

After You Read the Module

Review Key Terms
Match the key terms on the left with the definitions on the right.

_____ 1. Fossil fuels	a. A graph that represents oil use and projects both when world oil production will reach a maximum and when world oil will be depleted.
_____ 2. Nonrenewable energy resource	b. The amount of energy we get out of an energy source for every unit of energy expended on its production.
_____ 3. Renewable energy resources	c. Energy sources gathered by individuals for their own immediate needs including straw, sticks, and animal dung.
_____ 4. Potentially renewable	d. Energy sources that are bought and sold, such as coal, oil, and natural gas.
_____ 5. Nondepletable	e. The point at which oil extraction and use would increase steadily until roughly half the supply had been used up.
_____ 6. Commercial energy sources	f. An energy source that cannot be used up.
_____ 7. Subsistence energy sources	g. An energy source that can be regenerated indefinitely as long as it is not overharvested.
_____ 8. Energy intensity	h. The ratio of the amount of energy expended in the form you want to the total amount of energy that is introduced into the system.
_____ 9. Fossil fuel combustion	i. The energy use per unit of gross domestic product (GDP).
_____ 10. Hubbert curve	j. An energy source with a finite supply, primarily fossil fuels and nuclear fuels.
_____ 11. Peak oil	k. Fuels derived from biological material that became fossilized millions of years ago.
_____ 12. Energy conservation	l. The chemical reaction between any fossil fuel and oxygen resulting in the production of carbon dioxide, water, and the release of energy.
_____ 13. Energy efficiency	m. Sources of energy that are infinite.
_____ 14. Energy return on energy investment (EROEI)	n. Methods for finding and implementing ways to use less energy.

MODULE 36: Fuel Types and Uses

Before You Read the Module

Focus on Learning Goals

Use the module learning goals to guide your reading. On a separate piece of paper, write down each goal and take notes to help you meet each learning goal. After studying this module, you should be able to:

- 36-1 identify whether the Sun is the ultimate source of many fuels.
- 36-2 describe the major fuel types and how they are used.
- 36-3 identify other uses of fossil fuel.
- 36-4 describe electricity generation and cogeneration.

Key Terms

Biofuel	Lignite	Energy carrier
Modern carbon	Bituminous coal (Asphalt)	Combined cycle
Fossil carbon	Anthracite (hard coal)	Capacity
Carbon neutral	Natural gas	Capacity factor
Coal	Crude oil	Cogeneration (Combined
Peat	Tar sands (Oil sands)	heat and power)

While You Read the Module

Answer the following questions as you read. Use a separate sheet of paper if necessary.

Module 36: Fuel Types and Uses

The Sun is the ultimate source of many of the fuels we use

1. Figure 36.1: List the types of energy that have the Sun as their source and what role the Sun played in their production.

2. Figure 36.1: List the types of energy that are not produced from the Sun.

3. Identify the types of biomass that can be used as fuel.

4. Define biofuel.

Modern Carbon versus Fossil Carbon

5. Compare the age of carbon found in corn to carbon found in wood.

6. Define modern carbon.

7. Define fossil carbon and identify the age of fossil carbon.

8. Describe how burning fossil carbon could cause rapid increases in atmospheric carbon dioxide concentrations.

9. Explain how current biomass used or burned should be carbon neutral.

10. Define carbon neutral.

Each fuel has specific optimal applications

Wood

11. How is wood used worldwide?

12. Explain how cutting trees for fuel can be sustainable. Is current wood use sustainable?

13. Why is charcoal a more desirable energy source than wood?

Coal and Peat

14. Define coal.

15. Define peat.

16. List the three types of coal.

17. Define lignite.

18. Define bituminous coal.

19. Define anthracite.

20. Figure 36.2: Identify and describe the four stages of coal formation.

Natural Gas

21. Define natural gas.

22. Identify the gases included in natural gas.

23. List the uses of natural gas.

24. Describe liquefied petroleum gas (LPG) and its uses.

Crude Oil

25. Define crude oil.

26. What substances are equivalent to crude oil?

27. Identify what crude oil can be further refined to.

28. How many oil refineries are located in the United States?

29. Identify the quantity of oil contained in a barrel of oil.

30. Figure 36.3: Describe the crude oil distillation process.

Tar Sands

31. Define tar sands.

32. Describe bitumen.

33. Describe the natural movement of bitumen. How is bitumen accessed by humans?

34. Describe and explain the detrimental effects of tar sand extraction on the environment.

Fossil fuels have specialized uses for motor vehicles and electricity

Hot Water Heaters

35. Describe how an electric hot water heater heats water.

36. Figure 36.4: Explain the efficiency differences between an electric hot water heater and a natural gas water heater.

37. Describe why the overall efficiency of an electric hot water heater might be lower than a gas-powered water heater.

38. What is the relatively new and more efficient hot water heater?

Fossil Fuel Choices and Transportation

39. What percent of energy in 2020 was used for transportation in the United States?

40. Table 36.1: List the forms of transportation shown in the table from most efficient to least efficient.

41. Identify the cars preferred in the United States. Explain the concern with these cars.

42. List the percentage of new car sales that were electric vehicles in the United States, China, and Europe in 2020.

43. How can a self-driving car be more efficient?

Generation and cogeneration convert fuels to electricity

44. Define energy carrier.

45. How much energy is consumed in the United States to generate electricity and how much of the total consumed is available for end use?

The Process of Electricity Generation

46. Complete this statement; "The energy source that entails the fewest conversions…"

Efficiency of Electricity Generation

47. Identify the efficiency of coal-burning power plants.

48. Define combined cycle.

49. Define capacity.

50. Identify the capacity of a typical power plant in the United States.

51. Describe why a power plant would shut down.

52. Define capacity factor.

Cogeneration

53. Define cogeneration.

54. Describe how cogeneration is used to produce greater efficiencies.

55. Figure 36.5: Identify how cogeneration is used to heat or cool a building.

56. Figure 36.6: List the top four fuels used in 2020 for electricity generation in the United States.

Visual Representation 6 Transportation Decisions

57. Describe how an electric bicycle assists a rider.

58. Describe a disadvantage of an electric bicycle.

59. Describe the different types of public transportation discussed in different countries.

60. Make a chart listing the types of personal vehicles shown and how each is powered or recharged.

Practice the Math: Efficiency of Travel

Read "Do the Math: Efficiency of Travel" on page 417. Try "Your Turn." For more math practice, do the following problem. Remember to show your work. Use a separate sheet of paper if necessary.

If you could carpool with two other people from San Diego, California to Orlando, Florida, (a distance of roughly 3,910 km) what would the energy expenditure be per person? Would this be better than the other modes of transportation? Use the following information from page 422:

Air   2.1 MJ/passenger-kilometer × 3,910 km/trip = 8,211 MJ/passenger trip
Car   3.6 MJ/passenger-kilometer × 3,910 km/trip = 14,076 MJ/passenger trip
Train 1.1 MJ/passenger-kilometer × 3,910 km/trip = 4,301 MJ/passenger trip
Bus   1.7 MJ/passenger-kilometer × 3,910 km/trip = 6,647 MJ/passenger trip

Practice the Math: Calculating Energy Supply

Read "Do the Math: Calculating Energy Supply," on page 422. Try "Your Turn." For more math practice, do the following problem. Remember to show your work. Use a separate sheet of paper if necessary.

According to the U.S. Department of Energy, a typical home in the United States uses approximately 900 kWh of electricity per month. On an annual basis, this is

900 kWh/month × 12 months/year = 10,800 kWh/year

During summer months in Alaska, some homes don't run an air conditioner very often. How many homes can the same power plant support if average electricity usage in Alaska decreases to 600 kWh/ month during summer months?

After You Read the Module

Review Key Terms
Match the key terms on the left with the definitions on the right.

_____ 1. Biofuel

_____ 2. Modern carbon

_____ 3. Fossil carbon

_____ 4. Carbon neutral

_____ 5. Coal

_____ 6. Peat

_____ 7. Lignite

_____ 8. Bituminous coal (Asphalt)

_____ 9. Anthracite (hard coal)

_____ 10. Natural gas

_____ 11. Crude oil

_____ 12. Tar sands (Oil sands)

_____ 13. Energy carrier

_____ 14. Combined cycle

_____ 15. Capacity

_____ 16. Capacity factor

_____ 17. Cogeneration (Combined heat and power)

a. A brown coal that is a soft sedimentary rock that sometimes shows traces of plant structure; it typically contains 60 to 70 percent carbon.

b. A relatively clean fossil fuel containing 80 to 95 percent methane (CH_4) and 5 to 20 percent ethane, propane, and butane.

c. A precursor to coal, made up of partly decomposed organic material, including mosses.

d. An activity that does not change atmospheric CO_2 concentrations.

e. A black or dark brown coal that contains bitumen. It typically contains up t 80 percent carbon.

f. A liquid fuel such as ethanol or biodiesel created from processed or refined biomass.

g. It contains greater than 90 percent carbon. It has the highest quantity of energy per volume of coal and the fewest impurities.

h. The fraction of time a power plant operates during a year.

i. Old carbon contained in fossil fuels.

j. A solid fuel formed primarily from the remains of trees, ferns, and other plant materials that were preserved 280 million to 360 million years ago.

k. A mixture of hydrocarbons such as oil, gasoline, kerosene as well as water and sulfur that exists in a liquid state underground, and when brought to the surface.

l. A feature in some natural gas–fired power plants that uses both a steam turbine to generate electricity and a separate turbine that is powered by the exhaust gases from natural gas combustion to turn another turbine to generate electricity.

m. An energy source such as electricity that can move and deliver energy in a convenient, usable form to end users.

n. The use of a fuel to both generate electricity and deliver heat to a building or industrial process.

o. Carbon in biomass that was recently in the atmosphere.

p. The maximum electrical output of something such as a power plant.

q. A mixture of hydrocarbons such as oil, gasoline, kerosene as well as water and sulfur that exists in a liquid state underground, and when brought to the surface.

MODULE 37: Distribution of Natural Energy Resources: Fossil Fuels

Before You Read the Module

Focus on Learning Goals

Use the module learning goals to guide your reading. On a separate piece of paper, write down each goal and take notes to help you meet each learning goal. After studying this module, you should be able to:

- 37-1 explain why fossil fuels and ores are found only in certain locations.
- 37-2 describe the advantages and disadvantages of fossil fuels including oil extraction and fracking.
- 37-3 describe how fossil fuels are used for electricity generation.

Key Terms

Fracking
Volatile organic compounds (VOCs)
Turbine

Electrical grid
Energy quality

While You Read the Module

Answer the following questions as you read. Use a separate sheet of paper if necessary.

Module 37: Distribution of Natural Energy Resources: Fossil Fuels

Fossil fuel and ore distribution around the globe depends on the geology of the region

1. List where the largest coal reserves are located.

2. List the countries that are currently producing the greatest amounts of coal.

3. Figure 37.1: Explain why a flame is burning next to an oil well.

4. Identify the amount of petroleum and petroleum products used per day in the United States.

5. Identify other raw materials developed from petroleum.

6. List the top oil producing countries.

Fossil fuels have many advantages and disadvantages

Advantages of Coal

7. Explain why coal is used.

8. List the advantages of coal.

Disadvantages of Coal

9. Describe the disadvantages of surface mining and subsurface mining.

10. Identify the impurities coal contains.

11. Describe the chemical spill in West Virginia in 2014.

12. How many coal mines are in the United States? Describe the change that has occurred in the last 5-10 years.

13. Figure 37.2: Describe the disaster shown in this figure and explain its environmental impact.

14. Explain the significance of the carbon dioxide released from coal.

Advantages of Oil

15. List the advantages of oil.

Disadvantages of Oil

16. Identify trace elements found in oil.

17. Identify the possible ways oil can spill into the environment.

18. Describe the cause of the oil spilled during the Persian Gulf War in 1991 and name the amount that was spilled.

19. Explain other means of oil entering the marine waterways.

20. Describe the concerns about building an oil pipeline through the North Slope of Alaska.

21. Figure 37.3: Explain the Lac-Mégantic railway accident.

22. Describe the Arctic National Wildlife Refuge (ANWR).

23. Contrast the views of those for and against drilling for oil in the ANWR.

24. Explain the current lease status of the ANWR.

25. Figure 37.4(a): Describe the "ten-o-two" area.

26. Describe the effects of oil extraction in Nigeria and other developing countries.

Advantages of Natural Gas

27. Identify the advantages of natural gas.

Disadvantages of Natural Gas

28. Explain why escaped methane is a concern.

29. Figure 37.5: Identify possible ecosystem disturbances shown in the picture.

30. Define fracking.

31. Describe the benefits of fracking.

32. How have the energy sources for electricity generation in the United States changed since 1980? Compare 1980 with 2020.

33. Name FOUR examples of negative consequences of fracking.

34. Figure 37.6: Describe how fracking works.

35. Define volatile organic compounds (VOCs).

36. Describe fugitive gas and the controversial concern.

Fuel is converted to electricity and releases carbon dioxide and heat energy

37. Figure 37.7: List the steps in the process of using coal to generate electricity.

38. Define turbine.

39. Define electrical grid.

40. Figure 37.8: What is the calculation of energy efficiency of coal from a power plant to a light bulb? What do the red arrows indicate?

Energy Quality

41. Define energy quality.

42. Describe the characteristics of a high-quality energy source.

43. Identify one example of a high-quality energy source and one example of a low-quality energy source.

Practice the Math: Comparing the Efficiencies of Two Systems

Read "Do the Math: Comparing the Efficiencies of Two Systems" on page 432. Try "Your Turn." For more math practice, do the following problem. Remember to show your work. Use a separate sheet of paper if necessary.

Electric water heaters are close to 100 percent efficient at the source, but the electricity originates at a coal power plant (35 percent efficient) and travels through transmission lines (90 percent efficient). Natural gas water heaters are approximately 50 percent efficient at the source and may lose up to 15 percent of fuel through transmission. Calculate the overall efficiencies of the two systems.

After You Read the Module

Review Key Terms
Match the key terms on the left with the definitions on the right.

_____	1. Fracking	a.	A type of organic compound air pollutants that evaporate at typical atmospheric temperatures.
_____	2. Volatile organic compounds (VOCs)	b.	The ease with which an energy source can be used to do work.
_____	3. Turbine	c.	A device that can be turned by water, steam, or wind to produce power such as electricity.
_____	4. Electrical grid	d.	A network of interconnected transmission lines.
_____	5. Energy quality	e.	Short for hydraulic fracturing, a method of oil and gas extraction that uses high-pressure fluids to force open existing cracks in rocks deep underground.

MODULE 38: Nuclear Power

Before You Read the Module

Focus on Learning Goals
Use the module learning goals to guide your reading. On a separate piece of paper, write down each goal and take notes to help you meet each learning goal. After studying this module, you should be able to:
- 38-1 explain how nuclear energy is used to generate electricity.
- 38-2 describe advantages and disadvantages of nuclear power.
- 38-3 explain radioactivity and radioactive waste.
- 38-4 identify the three major nuclear accidents.

Key Terms

Nuclear power	Control rod	Becquerel (Bq)
Radioactivity	Radioactive decay	Curie
Fission	Half-life	
Fuel rod	Radioactive waste	

While You Read the Module
Answer the following questions as you read. Use a separate sheet of paper if necessary.

Module 38: Nuclear Power

1. Define nuclear power.

Nuclear reactors use fission to generate electricity

2. Identify the fuel source used for nuclear energy.

3. Define radioactivity.

4. Define fission.

5. Figure 38.1: Explain why nuclear fission occurs through a chain reaction.

6. Compare and contrast the energy produced from coal with the energy produced from uranium-235.

7. Define fuel rod.

8. Figure 38.2: Does the water surrounding the control rods (green color) mix with water from the steam generator (blue and orange colors)? Explain.

9. Identify the type of nuclear power plants used in the United States.

10. Define control rod.

11. Describe a meltdown of a nuclear power plant.

Concentrating the Uranium Ore

12. Identify how much uranium ore is needed to produce nuclear fuel.

13. What percentage of uranium-235 must suitable nuclear fuel contain?

Nuclear power has advantages and disadvantages

14. Identify the advantages of nuclear energy.

15. List three countries with nuclear energy.

16. Explain why generating nuclear energy in the United States has become so expensive.

17. Identify the reasons for public protests.

18. Explain the recent resurgence of nuclear energy in the United States.

Nuclear power depends on radioactivity but as a result, it generates radioactive waste

Radioactive Isotopes Undergo Radioactive Decay

19. Define radioactive decay.

20. Define half-life. Describe how half-life is a useful scientific tool.

Radioactive Waste: The By-Product of Electricity Generated from Nuclear Power

21. Define radioactive waste.

22. Identify and describe the three types of radioactive waste.

23. Describe the effects on humans from the different types of radioactive waste.

Measuring Half-Lives

24. Identify the half-life of uranium-235.

25. Define a becquerel (Bq).

26. Define a curie.

27. Describe how the curies change with the passing of each half-life.

28. Identify how long the spent fuel rods are a threat to human health.

29. Identify where and how spent fuel rods are stored.

30. Identify all challenges in disposing of radioactive waste.

31. Explain the criteria for safe storage of radioactive waste.

32. Describe the location of a possible long-term storage for spent nuclear fuel.

33. Describe the future of Yucca Mountain as a nuclear fuel storage area.

Three Mile Island, Chernobyl, and Fukushima are the three nuclear accidents

34. Describe where the Three Mile Island nuclear power plant is located and when the accident occurred.

35. Describe the cause of the accident at Three Mile Island.

36. Explain the results of the partial meltdown.

37. Describe what happened to the facility as the result of the accident.

38. Identify where and when the nuclear accident occurred in Ukraine.

39. Explain the cause of the nuclear accident in Chernobyl.

40. Describe a "runaway" reaction.

41. List the detrimental effects from the nuclear accident.

42. Explain what led to the Japan nuclear power plant accident.

43. Describe the results to the power plant after the tsunami caused flooding.

44. Identify the human impacts from the Japan nuclear accident.

45. Describe Japan's response after the accident.

Nuclear Power Compared with Other Fuels

46. Describe the status of new nuclear power plants in the United States.

47. Table 38.1: List the energy types from highest to lowest energy return on energy investment (EROEI). Identify the rate for each.

48. Table 38.1: List the lowest to highest cost for electricity (cents/kWh). Identify the rate for each.

Practice the Math: Calculating Half-Lives

Read "Do the Math: Calculating Half-Lives" on page 439. Try "Your Turn." For more math practice, do the following problem. Remember to show your work. Use a separate sheet of paper if necessary.

Uranium-235 has a half-life of 700 million years. How many years will it take uranium-235 to decay to ⅛ of its original mass?

After You Read the Module

Review Key Terms

Match the key terms on the left with the definitions on the right.

_____   1. Nuclear power

      a. A measurement of the rate at which a sample of radioactive material decays; 1 Bq is equal to the decay of one atom per second.

_____   2. Radioactivity

      b. When a parent radioactive isotope emits alpha or beta particles or gamma rays.

_____   3. Fission

      c. Electricity generated from the nuclear energy contained in nuclear fuel.

_____   4. Fuel rod

      d. The time it takes for one-half of the original radioactive parent atoms to decay.

_____   5. Control rod

      e. A cylindrical tube that encloses nuclear fuel within a nuclear reactor.

_____   6. Radioactive decay

      f. Nuclear fuel that can no longer produce enough heat to be useful in a power plant but continues to emit radioactivity.

_____   7. Half-life

      g. The emission of ionizing radiation or particles caused by the spontaneous disintegration of atomic nuclei.

_____   8. Radioactive waste

      h. A unit of measure for radiation, a curie is 37 billion decays per second.

_____   9. Becquerel (Bq)

      i. A cylindrical device inserted between the fuel rods in a nuclear reactor to absorb excess neutrons and slow or stop the fission reaction.

_____   10. Curie

      j. A nuclear reaction in which a neutron strikes a relatively large atomic nucleus, which then splits into two or more parts, releasing additional neutrons and energy in the form of heat.

MODULE 39: Biomass and Solar Energies and Hydroelectric Power

Before You Read the Module

Focus on Learning Goals

Use the module learning goals to guide your reading. On a separate piece of paper, write down each goal and take notes to help you meet each learning goal. After studying this module, you should be able to:
- 39-1 explain positive and negative consequences of biomass energy resources.
- 39-2 describe the various types of solar energy systems.
- 39-3 explain how hydroelectric power is generated and describe its environmental effects.

Key Terms

Biomass	Carbon dioxide	Photovoltaic solar cells
Charcoal	Biofuel	Hydroelectricity
Particulates (Particulate	Ethanol	Water impoundment
matter; Soot)	Biodiesel	Run-of-the-river
Carbon monoxide	Passive solar	Tidal energy
Nitrogen oxides	Active solar energy	Siltation

While You Read the Module

Answer the following questions as you read. Use a separate sheet of paper if necessary.

Module 39: Biomass and Solar Energies and Hydroelectric Power

Biomass energy resources are derived from biological material and can displace fossil fuels

Solid Biomass: Wood, Charcoal, and Manure

1. Define biomass.

2. List the three most commonly used biomass resources.

3. Define charcoal.

4. Describe when biomass can be a renewable resource.

5. Identify possible detrimental effects of tree removal.

6. Explain why charcoal is a preferred source of energy over wood.

7. List the air pollutants commonly associated with burning wood and charcoal as a fuel for cooking.

8. Define particulates.

9. Define carbon monoxide.

10. Define nitrogen oxides.

11. Define carbon dioxide.

12. Figure 39.1(a): Describe a detrimental environmental effect of charcoal production.

13. Why is animal manure a fuel choice in developing nations?

14. Explain how burning manure can have detrimental effects.

15. Figure 39.2: Describe the pollution in the photograph.

Liquid Biofuels: Ethanol and Biodiesel

16. Define biofuel.

17. List the two types of biofuels.

18. Define ethanol. Identify the plants used to make ethanol.

19. Define biodiesel. Identify the plants used to make biodiesel.

20. Identify the top two world producers of ethanol.

21. Describe the difference between corn and sugarcane used for ethanol.

22. Describe gasohol and a benefit of this fuel.

23. Describe the disadvantages of ethanol.

24. Describe biodiesel B-20.

25. Explain what diesel engines can use biodiesel.

26. Identify where most biodiesel comes from in the United States.

27. Describe the reasons algae would be a beneficial source for producing biodiesel.

28. Explain why producing biodiesel from soybeans may be a poor environmental choice.

29. Describe SVO and what it is known as.

30. Figure 39.3: Identify two benefits of using recycled cooking oil as a biodiesel.

The energy of the Sun can be captured passively and actively

31. Figure 39.4: Identify the location of the United States that receives the greatest amount of daily solar radiation and the amount.

Passive Solar Heating

    32. Define passive solar.

    33. Identify the possible cooking temperatures for "box cookers" and describe the items that can be cooked.

    34. Describe the environmental and social benefits of solar ovens.

Active Solar Energy Technologies

    35. Define active solar energy and give examples of the equipment used.

    36. Identify solar water heating applications.

    37. Figure 39.6: Describe how solar water heating systems work.

    38. Explain the simplest solar water heating systems.

    39. Identify the backup to a solar water heating system and explain why this would be needed.

    40. Define photovoltaic solar cells.

    41. Figure 39.7a: Describe the steps that photovoltaic solar systems use to generate electricity.

    42. Explain the warranty period and efficiency of solar photovoltaic cells.

    43. Describe where extra electricity from a photovoltaic solar cell house goes if unused.

44. Explain how homes that are "off the grid" might use photovoltaic solar cells.

45. Identify other locations and their uses for photovoltaic solar cells.

46. Describe concentrating solar thermal (CST) systems.

47. Describe where concentrating solar thermal facilities are built and why.

48. Describe the drawbacks of concentrating solar thermal systems.

49. Figure 39.8: Describe where the Sun's energy is reflected to and how energy is generated.

Benefits and Drawbacks of Active Solar Energy Systems

50. Describe the benefits of active solar energy systems.

51. Describe the economic advantages to homeowners of installing solar energy systems.

52. Explain why photovoltaic solar panels are expensive.

53. Identify and describe ways to encourage the growth of active solar energy.

54. Identify the environmental and financial costs of photovoltaic solar cells.

The kinetic energy of water can generate electricity but there are consequences

55. Define hydroelectricity.

56. Identify the five states that generate the most hydroelectricity.

57. List the top hydroelectricity producers in the world, from greatest to least.

Methods of Generating Hydroelectricity

58. Describe the process of producing electricity from water.

59. Describe what determines the amount of energy being produced with hydroelectricity.

60. Define water impoundment.

61. Figure 39.9: Describe how operators of dams change the generation of electricity.

62. Identify the largest hydroelectric impoundment dam in the United States and the world.

63. Describe the difference in electricity generation between the Grand Coulee Dam and the Three Gorges Dam.

64. Define run-of-the-river.

65. List advantages of run-of-the-river systems.

66. List disadvantages of run-of-the-river systems.

67. Define tidal energy.

68. Describe the disadvantages of tidal energy.

Hydroelectricity and Sustainability

69. List the benefits of hydroelectric systems.

70. Figure 39.10: Explain the "bathtub" ring at the Glen Canyon Dam.

71. List and describe negative environmental consequences of hydroelectric systems.

72. Define siltation.

73. Figure 39.11: Explain the changes that occurred when the Marmot Dam was removed.

Practice the Math: Calculating Home PV Generation and Capacity Factor

Read "Do the Math: Calculating Home PV Generation and Capacity Factor" on page 451. Try "Your Turn." For more math practice do the following problems. Remember to show your work. Use a separate sheet of paper if necessary.

A home solar system has a small 1,150-watt solar array consisting of five 230-watt photovoltaic panels. In 2018, the array generated 1,250 kWh of electricity. Electricity in Georgia costs $0.09 per kWh.

(a) What is the capacity factor of this photovoltaic system?

(b) How much did the homeowner offset in electricity costs?

Review Key Terms
Match the key terms on the left with the definitions on the right.

_____   1. Biomass

_____   2. Charcoal

_____   3. Particulates (Particulate matter; Soot)

_____   4. Carbon monoxide

_____   5. Nitrogen oxides

_____   6. Carbon dioxide

_____   7. Biofuel

_____   8. Ethanol

_____   9. Biodiesel

_____   10. Passive solar

_____   11. Active solar energy

_____   12. Photovoltaic solar cells

_____   13. Hydroelectricity

_____   14. Water impoundment

_____   15. Run-of-the-river

_____   16. Tidal energy

_____   17. Siltation

a. Alcohol made by converting starches and sugars from plant material into alcohol and CO_2.

b. A by-product of combustion of any fuel in the atmosphere (which contains 78 percent nitrogen).

c. The storage of water in a reservoir behind a dam.

d. A use of technology that captures and stores the energy of sunlight with electrical equipment and devices.

e. Energy that comes from the movement of water driven by the gravitational pull of the Moon.

f. Liquid fuel created from processed or refined biomass.

g. Sediments from moving water that accumulate on the bottom of a reservoir.

h. A by-product of all combustion, carbon dioxide from biofuels contains modern carbon from woody material, rather than fossil carbon from fossil fuels.

i. Solid or liquid particles suspended in the air.

j. A diesel substitute produced by extracting and chemically altering oil from plants.

k. Woody material that has been heated in the absence of oxygen so that water and some volatile compounds are driven off.

l. A use of energy from the Sun that takes advantage of solar radiation without active technology.

m. Electricity generated by the kinetic energy of moving water.

n. A colorless, odorless gas that is formed during incomplete combustion of most materials.

o. Biological material that has mass.

p. A use of energy from the Sun as light, not heat, and converting it directly into electricity

q. Hydroelectricity generation in which water is retained behind a low, small dam or no dam.

MODULE 40: Geothermal Energy and Hydrogen Fuel Cells

Before You Read the Module

Focus on Learning Goals
Use the module learning goals to guide your reading. On a separate piece of paper, write down each goal and take notes to help you meet each learning goal. After studying this module, you should be able to:
- 40-1 explain how humans harness geothermal energy.
- 40-2 describe the hydrogen fuel cell and its potential for providing electricity.

Key Terms

Geothermal energy   Fuel cell
Ground source heat pump   Electrolysis

While You Read the Module
Answer the following questions as you read. Use a separate sheet of paper if necessary.

Module 40: Geothermal Energy and Hydrogen Fuel Cells

Earth's internal heat is transferred to water that we use for heating and electricity generation

1. Define geothermal energy.

2. Identify how humans can access geothermal energy.

3. Identify the countries that can access geothermal resources.

Harvesting Geothermal Energy

4. Identify the percent of homes and businesses in Iceland that are heated by geothermal energy.

5. Describe how geothermal energy can produce electricity.

6. Explain how geothermal energy could be depleted.

7. Compare the amount of electricity produced using geothermal energy in Iceland to that produced in the United States.

8. Figure 40.2: Describe what regions in the United States have the greatest amounts of geothermal resources.

9. List the states with geothermal power plants in the United States.

10. Identify a possible drawback of geothermal power plants.

Ground Source Heat Pumps

11. Define ground source heat pumps.

12. Identify and describe Earth's temperature at 3 meters (10 feet) underground.

13. Identify the source of energy for ground source heat pumps.

14. Figure 40.3: Identify the equipment used to transfer heat to the house.

15. Describe how a ground source heat pump works in the winter.

16. Describe how a ground source heat pump works in the summer.

17. Describe the benefits of ground source heats pumps.

18. Describe hot water heat pumps.

19. Identify similar systems to the hot water heat pumps.

20. Describe how a hot water heat pump is similar to a refrigerator and explain how the hot water heat pump works.

21. Describe the energy efficiency of hot water heat pumps.

22. Describe how states are incentivizing the use of hot water heat pumps.

Hydrogen fuel cells use hydrogen as an energy source and are almost pollution free

23. Define fuel cell.

24. Explain the difference between a fuel cell and a common battery.

25. Figure 40.4a: List the steps a fuel cell goes through to generate electricity.

26. Describe the challenges in providing hydrogen gas to the fuel cells.

27. Define electrolysis.

28. Describe how electrolysis could be sustainable.

29. Identify how many hydrogen fuel cells operate in the United States and the amount of energy that is produced.

30. Identify the percent efficiency of hydrogen fuel cells.

31. List and describe the disadvantages of hydrogen as a fuel.

32. Describe why hydrogen-fueled vehicles are considered sustainable and a viable energy alternative.

After You Read the Module

Review Key Terms
Match the key terms on the left with the definitions on the right.

_____	1. Geothermal energy	a. An electrical-chemical device that converts fuel, such as hydrogen, into an electrical current.
_____	2. Ground source heat pump	b. Heat energy that comes from the natural radioactive decay of elements deep within Earth.
_____	3. Fuel cell	c. The application of an electric current to water molecules to split them into hydrogen and oxygen.
_____	4. Electrolysis	d. A technology that transfers heat from the ground to a building.

MODULE 41: Wind Energy and Energy Conservation

Before You Read the Module

Focus on Learning Goals
Use the module learning goals to guide your reading. On a separate piece of paper, write down each goal and take notes to help you meet each learning goal. After studying this module, you should be able to:
- 41-1 describe the benefits and impacts of wind energy.
- 41-2 explain the methods of conserving energy.

Key Terms

Wind energy	Peak demand	Smart grid
Wind turbine	Passive solar design	Oxygenated fuel
Phantom loads	Thermal mass	Cellulosic ethanol

While You Read the Module
Answer the following questions as you read. Use a separate sheet of paper if necessary.

Module 41: Wind Energy and Energy Conservation

Wind energy is the most rapidly growing source of electricity

1. Define wind energy.

2. Describe how unequal heating of Earth by the Sun creates wind.

3. Figure 41.1: Describe the global growth of installed wind energy capacity from 2000 to 2020.

4. Figure 41.2: Compare and contrast installed and generated wind energy between China and Denmark.

5. List the five countries with the highest installed wind energy capacity.

6. List the states with the largest amount of generating capacity in the United States.

Generating Electricity from Wind

7. Define wind turbine.

8. Figure 41.3: Describe how a wind turbine generates electricity.

9. Explain how wind turbine design has changed and why.

10. Describe how much energy a wind turbine may produce.

11. Explain why offshore wind turbine locations are more desirable than land.

12. Identify the best locations for land wind turbines.

13. Describe wind farms or wind parks.

14. List countries with operating offshore wind parks.

15. Describe why the Cape Wind near-shore wind project in Massachusetts was cancelled.

16. Describe the Rhode Island Block Island Wind Farm project and the energy output.

Benefits of Wind Energy

17. List the advantages of wind energy.

Disadvantages of Wind Energy

18. List the disadvantages of wind energy.

19. Identify the methods used to reduce deaths of birds from wind turbines.

We can use less, and use different technologies to conserve energy

20. Table 41.1: Copy the table "Ways to conserve energy.

21. Define phantom loads. Identify examples.

22. List possible ways the government could implement energy conservation measures.

23. Identify when brownouts or blackouts occur.

24. Define peak demand.

25. Explain how electric companies keep up with peak demand.

26. Describe variable price structures for electricity.

27. Identify when you think peak demand would occur for your home, town, or city.

28. Explain how reducing electricity use by 100 kWh can actually conserve 300 kWh.

Efficiency

29. Identify and describe efficiencies of different types of light bulbs.

30. Describe the Energy Star program.

31. Identify and describe an example of an Energy Star appliance saving electricity. Identify the amount of money that could be saved by using Energy Star.

Sustainable Design

32. Figure 41.6: List energy-efficient features incorporated into the sustainable home design.

33. Describe how community planning can conserve energy.

34. Define passive solar design.

35. List and describe features of a passive solar design in the Northern Hemisphere.

36. Figure 41.7: Describe how the design of the house adjusts to the differences in winter and summer.

37. Define thermal mass.

38. Identify and describe high thermal mass and low thermal mass materials.

39. Describe a green roof.

40. Explain the energy benefits of a green roof.

41. Explain how the use of recycled building materials assists with energy conservation.

42. Identify and explain two types of recycled materials used in building and where, specifically, they are used.

43. Figure 41.8: Identify the sustainable features of the California Academy of Sciences building.

44. Describe two active technologies that reduce energy use in the California Academy of Sciences building.

Energy Summary and Synthesis

45. Table 41.2: List the renewable energy types from highest to lowest energy return on energy investment (EROEI). Identify the rate for each.

46. Table 38.1 (page 441) and Table 41.2: List the energy types from highest to lowest energy return on energy investment (EROEI) for all entries in both tables. Identify the rate for each.

47. Using renewable and non-renewable energy resources from question 2 identify the source of the most efficient and least efficient EROEI value.

48. Table 41.2: List the lowest to highest cost for electricity (cents/kWh). Identify the rate for each.

49. Table 38.1 and Table 41.2: Integrate both tables and list the lowest to highest cost for electricity (cents/kWh). Identify the rate for each.

50. From the previous question identify the source of the lowest and highest costs for electricity (cents/kWh) from renewable and non-renewable energy resources.

51. Identify what must be combined to have a sustainable energy strategy.

52. Explain why the delivery of electricity from renewable energy could be problematic.

53. Identify the percentage of electricity lost in electrical transmission lines.

54. Describe why batteries are not an adequate energy storage sources to solve the problem of energy loss.

55. Define smart grid.

56. Figure 41.9: Identify the "smart" appliances in the house. Identify the differences between the blue and yellow connections to the computer technology and smart appliances.

57. Describe how the "smart" dishwasher runs.

Pursuing Environmental Solutions

Energy from Innovation and Renewables

58. Describe William Kamkwamba's early life.

59. Explain how William built a windmill.

61. Explain how William used electricity from the windmill.

62. Explain the power source for the hospital and women's health pavilion in Kigutu Burundi.

63. Describe William's recent life and his current work.

Science Applied 6

Should Corn Become Fuel?

64. Describe the differences between the proponents and opponents to ethanol.

Does ethanol reduce air pollution?

65. Describe burning hydrocarbons under ideal conditions.

66. Define oxygenated fuel.

67. Does ethanol reduce air pollution? Explain.

Is ethanol neutral in the productions of greenhouse gases?

68. Describe why ethanol should be carbon neutral.

69. Describe the life cycle of ethanol production and the net increase of carbon.

70. List and describe other detrimental effects of growing corn for ethanol production.

Does ethanol provide a substantial return on energy investment?

71. Figure SA6.2: Identify the return on energy investment for ethanol.

72. Compare the energy output of ethanol to that of gasoline.

Does ethanol reduce our dependence on gasoline?

73. Identify a better option than ethanol for reducing gasoline consumption.

What are the unintended impacts of ethanol production?

74. Identify the unintended consequences of ethanol production.

Are there alternatives to corn ethanol?

75. Define cellulosic ethanol.

76. List possible sources of cellulosic ethanol.

77. Explain the drawbacks of using cellulosic ethanol as a possible replacement for corn ethanol.

78. Explain the positives of using cellulosic ethanol.

What's the bottom line?

79. List the detrimental effects of ethanol production.

80. List the positives of ethanol production.

Practice the Math: Energy Star

Read "Do the Math: Energy Star" on page 472. Try "Your Turn." For more math practice do the following problems. Remember to show your work. Use a separate sheet of paper if necessary.

You are about to purchase a new refrigerator. You can choose between an Energy Star model, which costs $300, and a non-Energy Star model, which costs $200. The cost of electricity is $0.10 per kilowatt hour (kWh), and you expect your refrigerator to run an average of 10 hours per day.

(a) The non-Energy Star model uses 0.5 kW per hour. How much will it cost you per year for electricity to run this model?

(b) If the Energy Star model uses 0.4 kW per hour, how much money would you save on your electric bill over 5 years by buying the efficient model?

Review Key Terms

Match the key terms on the left with the definitions on the right.

_____	1. Wind energy	a. Electrical demand by a device that draws electrical current, even when it is turned off.
_____	2. Wind turbine	b. Construction technique designed to take advantage of solar radiation without active technology.
_____	3. Phantom loads	c. Energy generated from the kinetic energy of moving air.
_____	4. Peak demand	d. An efficient, self-regulating electricity distribution network that accepts any source of electricity and distributes it automatically to end users.
_____	5. Passive solar design	e. The greatest quantity of energy used at any one time.
_____	6. Thermal mass	f. An ethanol derived from cellulose, the cell wall material in plants.
_____	7. Smart grid	g. A turbine that converts the kinetic energy of moving air into electricity.
_____	8. Oxygenated fuel	h. A property of a building material that allows it to maintain heat or cold.
_____	9. Cellulosic ethanol	i. A fuel with oxygen as part of the molecule.

UNIT 6 Review Exercises

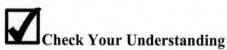Check Your Understanding

Review "Learning Goals Revisited" at the end of each module in Unit 1 of your textbook. Compare the notes you took while reading each module. Complete these exercises to review the unit. Use a separate sheet of paper if necessary.

1. Describe the types of nonrenewable and renewable energy resources. Include potentially renewable and nondepletable resources in your description.

2. Using the example of coal, describe losses of energy during extraction and use.

3. Describe modern carbon, fossil carbon, and carbon neutral. Give an example for each one.

4. Identify the different fuel types and their optimal application or uses.

5. Describe the formation environments of coal, oil, and natural gas.

6. List two advantages and two disadvantages of using nuclear power as an energy source.

7. Identify all the renewable energy resources powered by the Sun. Describe how each renewable energy resource is produced by the Sun's power.

8. Identify one advantage and disadvantage for each of the following energy resources:

Energy Resource	Advantage	Disadvantage
Liquid biofuels		
Solid biomass		
Photovoltaic solar cells		
Solar water heating systems		
Hydroelectricity		
Tidal energy		
Geothermal energy		
Wind energy		
Hydrogen fuel cell		

9. Identify three areas where you could conserve energy in your life. Give an example of each.

10. Explain the differences between active and passive solar.

Practice for Free-Response Questions

Complete this exercise to build and practice the skills you will need to answer free-response questions on the exam. Use a separate sheet of paper if necessary.

1. One negative consequence of nuclear power is that it generates waste, spent rods, that must be stored.
   (a) Identify where the waste is stored.

   (b) Explain the reasoning for the waste must be stored.

2. The California Academy of Sciences has a sustainable design that includes a living green roof. Identify two environmental benefits of a living green roof.

3. Solar radiation in the United States has geographic implications. Using Figure 39.4 page 449, explain why Arizona, Colorado, or New Mexico would be good locations for photovoltaic solar panels.

4. A family wishes to put up holiday lights for the month of December. They can choose between LED string lights or incandescent string lights. The LED lights use 4 watts. The incandescent string lights use 40 watts. The family will use 3 strands each and the lights will be lit for 10 hours for the entire month of December at a cost off $0.12 per kWh.
   (a) Determine the cost for the LED lights.

   (b) Determine the cost for the incandescent lights.

   (c) Explain which lights the family should use to save money.

5. Compare coal to natural gas and discuss their negative environmental effects when burning to produce electricity. Identify the better choice to produce electricity.

Unit 6 Multiple-Choice Review Exam

1. If the average person in the United States uses 10,000 watts of energy 24 hours a day for 365 days per year, how many kW of energy does the average person use in a year?
   (a) 1,000 kW
   (b) 3,650 kW
   (c) 3,650,000 kW
   (d) 87,600 kW

2. Developed countries contain ____ percent of the world's population and use more than ____ percent of the world's energy each year.
   (a) 5; 50
   (b) 20; 40
   (c) 5; 70
   (d) 1; 50

3. The energy source that is used most in the United States is
   (a) coal.
   (b) oil.
   (c) natural gas.
   (d) renewables.

4. Which is part of a coal-fired electricity generation plant?
   (a) pulverizer
   (b) distillation unit
   (c) control rods
   (d) compressor

5. Which is an example of cogeneration?
   (a) using steam that is generated to heat buildings to turn a turbine
   (b) using both coal and oil to create electricity
   (c) increasing nuclear power plants in major metropolitan areas
   (d) substituting anthracite coal for low grade lignite coal

6. Which list displays types of coal in order from most moisture and least heat to least moisture and most heat?
   (a) peat, lignite, bituminous, anthracite
   (b) peat, bituminous, lignite, anthracite
   (c) anthracite, bituminous, lignite, peat
   (d) bituminous, anthracite, lignite, peat

7. Which is correct?
   (a) Petroleum is found as a sedimentary rock layer.
   (b) Petroleum burns cleaner than natural gas.
   (c) Petroleum comes from the remains of ocean-dwelling phytoplankton that died millions of years ago.
   (d) Petroleum is considered a renewable resource.

8. Natural gas is generally found with
   (a) oil.
   (b) coal.
   (c) both oil and coal.
   (d) uranium.

9. Which electric vehicle will travel the farthest without recharging?
   (a) Hybrid electric vehicle (HEV)
   (b) Electric vehicle (EV)
   (c) Plug-in hybrid electric vehicles (PHEV)
   (d) Hydrogen fuel cell electric vehicle (FCEV)

10. 1 gram of Uranium-235 contains _____ times the energy of 1 gram of coal.
    (a) 1,000
    (b) 100,000-200,000
    (c) 2-3 million
    (d) 1 billion

11. Which is one difference between generating electricity with coal and generating electricity with nuclear energy?
    (a) Coal power generates steam and nuclear power does not.
    (b) Nuclear power uses fission to create heat to generate steam and coal does not.
    (c) Nuclear power produces more air pollution than coal.
    (d) Coal is much more energy efficient than nuclear energy.

12. If a material has a radioactivity level of 100 curies and has a half-life of 10 years, how many half-lives will have occurred after 100 years?
    (a) 1 half-life
    (b) 4 half-lives
    (c) 10 half-lives
    (d) 1,000 half-lives

13. Which is a nonrenewable energy source?
    (a) wind
    (b) wood
    (c) geothermal
    (d) nuclear

14. Planting a large, deciduous shade tree next to a southern window is an example of
    (a) active solar design.
    (b) photovoltaic systems.
    (c) energy star technology.
    (d) passive solar design.

15. The nuclear powerplant accident in Chernobyl, Ukraine was caused by
    (a) operators deliberately disconnecting the cooling system.
    (b) a tsunami started by a 9.0 earthquake.
    (c) accidentally removing the control rods.
    (d) inability for the aging facility to maintain the nuclear reactions.

16. Most ethanol produced in the United States comes from _____ while in Brazil most ethanol comes from _____.
    (a) corn; oil deposits
    (b) oil sands; sugarcane
    (c) corn; sugarcane
    (d) biomass; wood chips

17. Which is a benefit of a dam?
    (a) release of greenhouse gases
    (b) creates employment or jobs
    (b) disruption to the life cycle of aquatic species
    (d) leads to siltation

18. A photovoltaic cell is used to
    (a) capture sunlight and turn it into electricity.
    (b) burn biomass fuel.
    (c) generate passive solar energy.
    (d) generate wind power.

19. Which countries lead the world in the production of geothermal energy?
    (a) the United States, Denmark, and Finland
    (b) Iceland, the United States, and China
    (c) Indonesia, United States, and Philippines
    (d) Russia, the United States, and China

20. Which is a benefit of wind power?
    (a) It can only be on land.
    (b) It does not harm wildlife.
    (c) It is relatively expensive.
    (d) It is renewable.

21. A hydrogen fuel cell uses hydrogen and oxygen to make
    (a) only electricity.
    (b) carbon dioxide and electricity
    (c) only water
    (d) energy, electricity, and water

22. In the United States, the greatest amounts of geothermal resources are found in
    (a) Florida.
    (b) the West.
    (c) the Midwest.
    (d) the East Coast.

23. Which is a disadvantage of tidal energy?
    (a) It is extremely expensive to run.
    (b) It generates carbon dioxide, a greenhouse gas.
    (c) It is aesthetically displeasing.
    (d) It is geographically limited.

24. Some attention is focused on improving the electrical grid by
    (a) having more centralized power generation plants.
    (b) used more solar sources.
    (c) implementing a smart grid.
    (d) use more battery storage.

25. Ethanol is produced from
    (a) algae.
    (b) gasoline.
    (c) corn.
    (d) manure.

26. Which is an environmental problem associated with the use of nuclear power?
    (a) high levels of air pollution during operation
    (b) minimal but safe waste
    (c) the large amount of heat generated during operation
    (d) release of high levels of carbon dioxide

27. Which is an advantage of passive solar technology?
    (a) It is a depletable resource.
    (b) It is expensive to set up.
    (c) No technology is needed.
    (d) It is expensive to manufacture and install.

28. A sample of radioactive waste has a half-life of 20 years and an activity level of 4 curies. How many years will it take for the activity level of this sample be 0.5 curies?
    (a) 40 years
    (b) 60 years
    (c) 80 years
    (d) 100 years

29. Ethanol derived from cellulose is from
    (a) corn.
    (b) switchgrass.
    (c) algae.
    (d) manure.

UNIT 7

Atmospheric Pollution

Unit Summary

Air pollution is a pervasive problem that can be found both indoors and outdoors. Humans cause air pollution when they operate cars, factories, and other activities. Air pollution includes problems such as acid deposition and photochemical smog. Natural sources of air pollution include volcanoes. Legislation such as the Clean Air Act helps regulate air pollution, which can have benefits to the environment and to human health.

MODULES IN THIS UNIT

Module 42: Introduction to Air Pollution
Module 43: Photochemical Smog, Thermal Inversions, Atmospheric CO_2 and Particulates
Module 44: Indoor Air Pollutants
Module 45: Reduction of Air Pollutants
Module 46: Acid Rain and Noise Pollution

Unit Opening Case: *Cleaning Up in Chattanooga*

This case study discusses how the city of Chattanooga, Tennessee, has improved its air quality. One part of this case discusses how ground-level ozone concentrations came to be extremely high in the late 1990s. It goes on to discuss how the private and public sectors had to work together to help Chattanooga attain a lower ozone level. This case illustrates some of the challenges cities face in cleaning up pollution and the importance of cooperation between government and industry.

Do the Math

This unit contains the following "Do the Math" boxes to help prepare you for calculation questions you might encounter on the exam.
- "Unleaded versus Leaded Gasoline" (page 499)
- "Averaging Radon over Time" (page 515)
- "Calculating Annual Sulfur Reductions" (page 525)

To make sure you understand the concepts and techniques presented in these boxes, do the practice problems presented in the text as well as the additional "Practice the Math" problems that appear in Module 42, Module 44, and Module 45 of this study guide.

MODULE 42: Introduction to Air Pollution

Before You Read the Module

Focus on Learning Goals
Use the module learning goals to guide your reading. On a separate piece of paper, write down each goal and take notes to help you meet each learning goal. After studying this module, you should be able to:
- 42-1 identify the sources and effects of major air pollutants.
- 42-2 describe primary and secondary pollutants and how they change over time.

Key Terms

Air pollution	Photochemical oxidant	Hydrocarbons
Sulfur dioxide (SO$_2$)	Smog	Primary pollutant
Haze	Lead (Pb)	Secondary pollutant

While You Read the Module
Answer the following questions as you read. Use a separate sheet of paper if necessary.

Unit Opening Case: Cleaning Up in Chattanooga

1. Describe the recent air quality events in Chattanooga.

2. Describe the physical location of Chattanooga, Tennessee and the economic importance of the city.

3. Explain the air pollution in Chattanooga in 1957 and how geography was a factor.

4. Describe the 1969 Air Pollution Control Ordinance.

5. Explain how the recycling program contributed to improvement in air quality.

6. What are some negative consequences of being designated a "nonattainment area?"

7. How did Chattanooga avoid the nonattainment area status?

8. Explain how Chattanooga lowered the levels of ozone ahead of schedule.

9. Identify the rating the American Lung Association gave Chattanooga in 2017.

Module 42: Introduction to Air Pollution

10. Define air pollution.

Air pollution sources are widespread and their effects vary

11. Explain the idea that air pollution is a global system.

Classifying Pollutants

12. List the six criteria air pollutants identified under the Clean Air Act.

13. List other air pollutants not on the criteria list that are commonly measured and may have the potential to be harmful.

Coal and Oil

14. Describe the types of air pollutants emitted by coal, oil, and natural gas, and describe the relative amount of air pollutants released by each when combusted.

15. List the sources of sulfur dioxide as an air pollutant.

16. Table 42.1: Identify the effects and impacts of sulfur dioxide.

17. Why are nitrogen oxides referred to as NO_X?

18. Describe nitrogen oxide (NO) and nitrogen dioxide (NO_2).

19. Table 42.1: Identify the effects and impacts of nitrogen oxides.

20. Describe carbon monoxide and when it is formed.

21. Table 42.1: Identify the effects and impacts of carbon monoxide.

22. Describe carbon dioxide formation.

23. Describe particulate matter (PM), also called particulates, or particles.

24. List common sources of particulate matter.

25. Table 42.1: Identify the effects and impacts of particulate matter.

26. List sources of particulate matter.

27. Define haze.

28. Define photochemical oxidants.

29. Table 42.1: Identify the effects and impacts of ozone.

30. Define smog.

31. Define lead (Pb).

32. Table 42.1: List the human derived sources of lead (Pb).

33. Explain why lead (Pb) compounds were added to gasoline and describe the environmental impact.

34. Explain why people living in old buildings are more likely to be exposed to lead. What effect does lead exposure have on humans?

35. Identify the origins of mercury (Hg) and possible detrimental effects.

36. Describe VOCs.

37. Define hydrocarbons.

38. List the sources of volatile organic compounds.

39. How do VOCs have the potential to be harmful?

40. When do hydrocarbons become air pollutants?

Primary and secondary pollutants have decreased over time

Primary Pollutants

41. Define primary pollutant.

42. List the primary pollutants.

Secondary Pollutants

43. Define secondary pollutants.

44. Figure 42.2: List the secondary pollutants.

45. How can society control secondary pollutants?

Air pollution sources can be identified and have decreased over time

46. List both mobile sources and nonroad mobile sources of carbon monoxide and nitrogen oxides.

47. Describe sources of anthropogenic sulfur dioxide and particulate matter.

48. Figure 42.3: For each graph, list the greatest emission source of each criteria air pollutant and the percentage it emits.

49. Describe National Ambient Air Quality Standards.

50. When does a locality become subject to NAAQS penalties?

51. Figure 42.4: Describe the trends of air pollutants shown on the graph.

52. Figure 42.4: Identify how much lead has decreased.

53. Explain why lead has decreased so much in the last 25 years.

54. List three countries other than the United States that have detrimental emissions levels.

Practice the Math: Unleaded Versus Leaded Gasoline

Read "Do the Math: Unleaded Versus Leaded Gasoline" on page 499. Try "Your Turn." For more math practice, do the following problems. Remember to show your work. Use a separate sheet of paper if necessary.

While lead was phased out of gasoline for on-road vehicles by the Clean Air Act of 1970, it was still used in off-road vehicles such as airplanes and accounts for approximately half of atmospheric lead pollution in America. Like leaded automobile gasoline, leaded aviation fuel contains roughly 0.5 g of lead per liter. In response to pressure from people with environmental concerns, a low-lead aviation fuel was developed. Low-lead aviation fuel contains 20 percent less lead than the original fuel.

(a) How much lead would a private plane burning 5,500 liters of leaded aviation fuel per year emit in one year?

(b) How much lead does the low-lead aviation fuel contain per liter?

(c) How much lead would a private plane burning 5,500 liters of low-lead aviation fuel per year emit in one year?

After You Read the Module

Review Key Terms

Match the key terms on the left with the definitions on the right.

_____   1. Air pollution

a. Reduced visibility.

_____   2. Sulfur dioxide (SO₂)

b. A trace metal that occurs naturally in rocks and soils, is present in small concentrations in coal and oil and is a neurotoxin.

_____   3. Haze

c. A class of air pollutants formed as a result of sunlight acting on chemical compounds such as nitrogen oxides and sulfur dioxide.

_____   4. Photochemical oxidant

d. Pollutant compounds that contain carbon-hydrogen bonds, such as gasoline and other fossil fuels, lighter fluid, dry-cleaning fluid, oil-based paints, and perfumes.

_____   5. Smog

e. A corrosive gas that comes primarily from combustion of fuels such as coal and oil, including diesel fuel from trucks.

_____   6. Lead (Pb)

f. A primary pollutant that has undergone transformation in the presence of sunlight, water, oxygen, or other compounds.

_____   7. Hydrocarbons

g. A type of air pollution that is a mixture of oxidants and particulate matter.

_____   8. Primary pollutant

h. A polluting compound that comes directly out of a smokestack, exhaust pipe, or natural emission source.

_____   9. Secondary pollutant

i. The introduction of chemicals, particulate matter, or microorganisms into the atmosphere at concentrations high enough to harm plants, animals, and materials such as buildings, or to alter ecosystems.

MODULE 43: Photochemical Smog, Thermal Inversions, Atmospheric CO₂ and Particulates

Before You Read the Module

Focus on Learning Goals

Use the module learning goals to guide your reading. On a separate piece of paper, write down each goal and take notes to help you meet each learning goal. After studying this module, you should be able to:

- 43-1 explain photochemical smog and how to reduce it.
- 43-2 describe thermal inversions and how they relate to air pollution.
- 43-3 describe the natural sources of particulates and CO_2.

Key Terms

Evaporate	PM_{10}	Photochemical smog (Los
Sublimate	$PM_{2.5}$	Angeles–type smog;
Formaldehyde	Sulfurous smog (London-type	Brown smog)
Thermal inversion	smog; Gray smog; Industrial	
Inversion layer	smog)	

While You Read the Module

Answer the following questions as you read. Use a separate sheet of paper if necessary.

Module 43: Photochemical Smog, Thermal Inversions, Atmospheric CO₂ and Particulates

Photochemical smog is a complex combination of compounds and it can be reduced by decreasing emissions of its precursors

1. Describe the findings of the 2021 *State of the Air* report.

Chemistry of Ozone and Photochemical Smog Formation

2. How many categories of smog exist? What are they?

3. Define photochemical smog, also known as Los Angeles-type smog or brown smog.

4. Define sulfurous smog also known as London-type smog, gray smog, or industrial smog.

5. Identify economic harm caused by smog.

6. Figure 43.3: Draw and label the diagrams (a), (b), and (c).

7. Describe ozone formation during daylight as shown in Figure 43.3 (a).

8. Describe the natural process of ozone destruction.

9. Define evaporate.

10. Define sublimate.

11. Describe formaldehyde.

12. Describe ozone formation in the atmosphere when VOCs are added.

13. Explain how temperature influences formation of smog. Give an example.

14. Describe the human health impacts of secondary pollutants such as smog and photochemical oxidants.

Thermal inversions trap air pollutants close to Earth's surface

15. Figure 43.4: Describe the topography that is seen in both diagrams.

16. Define thermal inversions.

17. Define inversion layer.

18. Describe a severe pollution event during an inversion layer.

19. Describe the thermal inversion events of Tianjin.

Volcanoes, forest fires, respiration, and decomposition are natural processes responsible for particulate emissions and carbon dioxide

20. Describe processes in nature that are also sources of air pollution.

21. Describe how natural processes can lead to smog and photochemical oxidant pollution.

Natural Sources of Particulates

22. Describe natural sources of particulate matter.

23. Describe how particulate matter is measured and expressed in written format.

24. Compare and contrast PM_{10} and $PM_{2.5}$.

25. How can particulate matter impact photosynthesis?

Natural Sources of Carbon Dioxide

26. What is carbon dioxide a byproduct of?

Review Key Terms

Match the key terms on the left with the definitions on the right.

_____ 1. Photochemical smog (Los Angeles– type smog; Brown smog)

a. The process of converting from a solid to a gas or vapor.

_____ 2. Sulfurous smog (London-type smog; Gray smog; Industrial smog)

b. An atmospheric condition in which a relatively warm layer of air at mid-altitude covers a layer of cold, dense air below.

_____ 3. Evaporate

c. Smog that is dominated by oxidants such as ozone.

_____ 4. Sublimate

d. Smog dominated by sulfur dioxide, sulfate compounds, and particulate matter.

_____ 5. Formaldehyde

e. Particles of size 2.5 μm and smaller can travel further within the respiratory tract and are of even greater health concern.

_____ 6. Thermal inversion

f. The process of converting from a liquid to a gas or vapor.

_____ 7. Inversion layer

g. Particles smaller than 10 μm are called Particulate Matter-10 and are not filtered out by the nose and throat and can be deposited deep within the respiratory tract.

_____ 8. PM_{10}

h. The layer of warm air that traps emissions in a thermal inversion.

_____ 9. $PM_{2.5}$

i. A naturally occurring compound that is used as a preservative and as an adhesive in plywood and carpeting.

MODULE 44: Indoor Air Pollutants

Before You Read the Module

Focus on Learning Goals

Use the module learning goals to guide your reading. On a separate piece of paper, write down each goal and take notes to help you meet each learning goal. After studying this module, you should be able to:

- 44-1 identify the major indoor air pollutants and where they come from.
- 44-2 describe indoor air pollution differences in the developing and developed world.

Key Terms

Indoor air pollutants   Radon-222
Asbestos   Sick building syndrome

While You Read the Module

Answer the following questions as you read. Use a separate sheet of paper if necessary.

Module 44: Indoor Air Pollutants

1. What type of air pollution is the cause of more deaths and where does this occur?

Indoor air pollutants come from natural, human-made, and combustion sources

2. Define indoor air pollutants.

3. What kind of system is a house? Explain the detrimental effects of this system.

4. Identify air pollutants that are both indoor and outdoor air pollutants.

5. Describe sources of indoor air pollutants.

Carbon Monoxide

6. Identify and describe the most common cause of carbon monoxide in a building.

7. Explain the detrimental effect carbon monoxide has on the body.

8. Describe how carbon monoxide exposure most often occurs in the developed world.

9. Identify the best method to prevent death from carbon monoxide exposure.

10. Identify the sources of carbon monoxide in developing countries.

11. Why are children at higher risk for carbon monoxide poisoning?

Particulates

12. Describe sources of particulates indoors.

13. Describe human health problems associated with prolonged exposure to indoor particulate matter.

14. Describe what makes up dust.

15. Describe how particulate matter, pollen, and bacteria enter into buildings.

16. Describe mold and how it is introduced into buildings.

17. Identify human health impacts of exposure to mold.

Asbestos

18. Define asbestos.

19. Give an example of a use for asbestos.

20. Identify the health risks of asbestos and the individuals most at risk.

21. Explain why removal of asbestos is important. Describe the removal process.

Radon

22. Define radon-222.

23. Explain how humans are exposed to radon.

24. Identify the detrimental health effects of radon.

25. Explain how to remediate radon.

26. Figure 44.4: Identify the potential radon exposure for your location in the United States.

Volatile Organic Compounds

27. Identify where volatile organic compounds may originate from in homes.

28. Describe the uses of formaldehyde.

29. Describe the smell you may notice from formaldehyde. Give an example.

30. Discuss the health effects of formaldehyde over time.

31. Explain how homeowners can reduce their risk of exposure to VOCs.

Lead

32. Describe the greatest exposure to lead and why older buildings are more prone to lead contamination.

Other Pollutants

33. Describe how outdoor air pollution can become indoor air pollutants.

Indoor air pollution in the developing world comes mostly from indoor combustion while in the developed world tightly sealed buildings and VOCs are a bigger problem

34. How many deaths are due to air pollution each year?

Developing Countries

35. How do heating and cooking contribute to indoor air pollution in developing countries?

36. How do buildings differ between developing countries and developed countries?

Developed Countries

37. List and identify three factors that have led to indoor air pollution in developed countries.

38. Figure 44.5: Identify and explain two indoor air pollutants that could occur in your home.

39. Identify the reasons for sick building syndrome in newer buildings.

40. Define sick building syndrome.

41. List the EPA's four specific reasons for sick building syndrome.

Visual Representation 7: Sources of Atmospheric Pollution

42. Describe how lead, NO_x, ground level ozone, and sulfur dioxide contribute to outdoor air pollution and the resulting human health problems they cause.

43. Describe how asbestos, radon, particulate matter, carbon monoxide, and VOCs contribute to indoor air pollution and the resulting human health problems they cause.

Practice the Math: Averaging Radon over Time

Read "Do the Math: Averaging Radon over Time" on page 515. Try "Your Turn." For more math practice, do the following problems. Remember to show your work. Use a separate sheet of paper if necessary.

(a) A homeowner uses two different short-term radon collection devices to measure the same room over eight days. One device measured 3 pCi/L and the other device measured 6 pCi/L. What is the concentration of radon in the room? Is radon abatement recommended by the EPA?

(b) A homeowner uses two different short-term radon collection devices to measure the radon concentration. One device measured 1 pCi/L over an eight-day period and the other device measured 5 pCi/L over a four-day period. What is the concentration of radon in the room? Is radon abatement recommended by the EPA?

After You Read the Module

Review Key Terms
Match the key terms on the left with the definitions on the right.

_____ 1. Indoor air pollutants	a. Compounds that adversely affect the quality of air in buildings and structures.
_____ 2. Asbestos	b. A buildup of toxic pollutants in weatherized spaces, such as newer buildings in the developed world.
_____ 3. Radon-222	c. A long, thin, fibrous silicate mineral with insulating properties, which can cause cancer when inhaled.
_____ 4. Sick building syndrome	d. A radioactive gas that occurs naturally from the decay of uranium and is an indoor air pollutant.

MODULE 45: Reduction of Air Pollutants

Before You Read the Module

Focus on Learning Goals
Use the module learning goals to guide your reading. On a separate piece of paper, write down each goal and take notes to help you meet each learning goal. After studying this module, you should be able to:
- 45-1 explain how air pollutants can be reduced.
- 45-2 describe the various air pollution control devices and how they are used.

Key Terms

Vapor recovery nozzle   Scrubber
Catalytic converter   Electrostatic precipitator

While You Read the Module
Answer the following questions as you read. Use a separate sheet of paper if necessary.

Module 45: Reduction of Air Pollutants

Air pollution can be controlled by prevention, conservation, fuel switching, and regulatory practices

1. Identify the best way to decrease air pollution. Give an example.

2. What is an economic effect of using low-sulfur coal or oil?

Regulatory Practices

3. What levels of government have regulatory practices to reduce air pollution?

4. Define vapor recovery nozzle.

5. Describe air pollution measures a municipality might take other than vapor recovery nozzles.

6. Describe how Mexico City has tried to lower smog concentrations. Compare that with Beijing.

7. Discuss the development of buying and selling allowances of sulfur dioxide emissions for the Clean Air Act.

8. Identify the consequence of going over the sulfur allowance.

9. What if a company doesn't use all of its sulfur allowance?

10. How much has the total SO_2 emissions changed in the United States from 1982 to 2008?

Control devices convert pollutants to less harmful compounds or remove them from the exhaust stream

11. What are the two basic ways that pollution control devices function?

Converting Pollutants to Less Harmful Compounds

12. Define catalytic converter.

13. Figure 45.3: Describe how a catalytic converter works.

14. Why did the Clean Air Act need to address lead in addition to catalytic converters?

15. Explain if catalytic converters have been effective.

Removing Pollutants from Waste Streams

16. Define scrubber.

17. Where are scrubbers used?

18. Figure 45.4: Describe how a wet scrubber works.

19. Describe how gravitational settling removes particulate matter.

20. Describe baghouse filters.

21. Define electrostatic precipitator.

22. Describe how an electrostatic precipitator works.

23. Where are electrostatic precipitators typically used?

24. What is the tradeoff between burning temperatures and efficiency in industrial processes to reduce nitrogen oxide emissions?

Practice the Math: Calculating Annual Sulfur Reduction

Read "Do the Math: Calculating Annual Sulfur Reduction" on page 525. Try "Your Turn." For more math practice, do the following problem. Remember to show your work. Use a separate sheet of paper if necessary.

A researcher just published an article stating that CO_2 levels have increased by 8 ppm. If the level increased from 376 ppm to 384 ppm in 5 years, what is the percent change of these emissions?

After You Read the Module

Review Key Terms
Match the key terms on the left with the definitions on the right.

_____	1.	Vapor recovery nozzle	a. A device that removes particulate matter by using an electrical charge to make particles coalesce so they can be removed from the exhaust stream.
_____	2.	Catalytic converter	b. A device that uses a combination of lime and or water to separate and remove particles from industrial exhaust streams.
_____	3.	Scrubber	c. A device that uses chemicals to convert pollutants such as nitrogen oxide and carbon monoxide to nitrogen gas and carbon dioxide.
_____	4.	Electrostatic precipitator	d. A device that prevents VOCs from escaping into the atmosphere while a person is fueling their vehicle.

MODULE 46: Acid Rain and Noise Pollution

Before You Read the Module

Focus on Learning Goals

Use the module learning goals to guide your reading. On a separate piece of paper, write down each goal and take notes to help you meet each learning goal. After studying this module, you should be able to:

- 46-1 describe acid rain and its effects.
- 46-2 explain the sources and effects of noise pollution.

Key Terms

pH   Base   Noise pollution
Acid   Acid rain (Acid deposition)   decibel A scale (dbA)

While You Read the Module

Answer the following questions as you read. Use a separate sheet of paper if necessary.

Module 46: Acid Rain and Noise Pollution

Acid rain is acidified precipitation derived from air pollutants, which adversely affects terrestrial and aquatic systems and buildings

1. Define acid rain.

2. Define pH.

3. Define acid and identify the associated ion.

4. Define base and identify the associated ion.

5. Explain the pH scale.

6. Explain the phrase: "the pH scale is logarithmic." Name an example.

7. Figure 46.1: List the pH values of various water environments.

8. Describe how ocean acidification is occurring.

9. What happens when air pollution mixes with rainwater?

10. Define the two acids that compose acid rain or acid deposition.

11. What natural process can also cause acid deposition?

12. Identify the pH value of acid deposition.

How Acid Deposition Forms and Travels

13. Describe how acid deposition forms.

14. What are the reactants (and their sources) that lead to acid deposition?

15. Figure 46.2: Describe dissociation and the effects.

16. How far can secondary pollutants travel?

17. Describe how and when acid deposition in the United States began to decline.

18. Give an example of how acid deposition might form in one region and have detrimental effects in another region.

Effects of Acid Deposition on Terrestrial and Aquatic Ecosystems

19. Identify an example of direct acid deposition.

20. Identify an example of indirect acid deposition.

21. Describe how limestone bedrock alters the impacts of acid deposition.

22. Explain how acid deposition causes a mobilization of metals. Identify one environmental impact.

23. Describe the effects of acid deposition on aquatic ecosystems.

Effects of Acid Deposition on Humans and Infrastructure

24. List the detrimental effects of acid deposition on ecosystems and human made structures.

Noise pollution in the air can affect human health and behavior; noise pollution in water may interfere with animal communication

25. Define noise pollution.

Noise Pollution and Humans

26. Define the decibel A scale (db(A)).

27. List the decibels of various sounds and the impacts they have.

28. Describe the human health impacts of long-term exposure to noise.

29. Describe how noise pollution is an environmental justice issue.

Noise Pollution and Wildlife

30. Explain the impact that noise pollution can have on animals.

31. How does noise pollution impact aquatic ecosystems?

Pursuing Environmental Solutions

Better Indoor Cook Stoves

32. Describe the detrimental effects of some cooking fuels used by households in China, India, and sub-Saharan Africa.

33. List the benefits of more efficient cook stoves.

34. Why was the BioLite CampStove developed?

35. Identify the mechanism used on the camp stove that generates electricity.

36. Describe the features of the HomeStove.

37. Identify the other challenges ahead for more efficient cook stoves in developing countries.

Science Applied 7: Can Cap-and-Trade Reduce Pollution Emissions Equitably?

Can successes controlling sulfur translate to carbon emission reductions?

38. How did the CAAA of 1990 impact sulfur dioxide emissions?

39. What was one of the underlying reasons for the success of the cap-and-trade approach?

40. What were some additional benefits to the reduction in sulfur dioxide emissions?

What were the implications of sulfur cap-and-trade for the environmental justice movement?

41. Describe how certain communities were subjected to even higher rates of pollution than before the CAAA of 1990.

42. Describe arguments against the cap-and-trade system of the CAAA.

43. How do many wealthier communities prevent environmental injustice in their neighborhoods?

44. Why are sulfur dioxide and other pollutants from coal burning power plants a specific concern for environmental justice activists?

Could cap-and-trade be used to reduce carbon dioxide emissions in the United States?

45. Provide two reasons that carbon dioxide is an ideal pollutant to be managed through cap-and-trade.

After You Read the Module

Review Key Terms
Match the key terms on the left with the definitions on the right.

_____	1. pH	a. Unwanted sound that interferes with normal activities that is loud enough to cause health issues including hearing loss.
_____	2. Acid	b. The relative strength of acids and bases in a substance. It is a logarithmic scale, meaning that each number on the scale represents a change by a factor of 10.
_____	3. Base	c. A substance that contributes hydroxide ions to a solution.
_____	4. Acid rain (Acid deposition)	d. A logarithmic scale that measure both the loudness of sound and the frequency.
_____	5. Noise pollution	e. A substance that contributes hydrogen ions to a solution.
_____	6. decibel A scale (dbA)	f. Precipitation high in sulfuric acid and nitric acid.

UNIT 7 Review Exercises

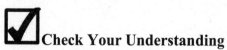Check Your Understanding

Review "Learning Goals Revisited" at the end of each module in Unit 1 of your textbook. Compare the notes you took while reading each module. Complete these exercises to review the unit. Use a separate sheet of paper if necessary.

1. Fill in the following chart.

Compound	Effects
Sulfur dioxide	
Nitrogen dioxide	
Carbon monoxide	
Particulate matter	
Lead	
Ozone	
VOCs	
Mercury	
Carbon dioxide	

2. Describe the formation of photochemical smog.

3. Explain the process of thermal inversion and its impact on air quality.

4. Differentiate between indoor air pollution found in developed countries versus developing countries.

5. Compare and contrast electrostatic precipitators and scrubbers.

6. Describe the causes and effects of acid deposition.

7. Describe the acute and chronic effects of noise pollution on humans.

Practice for Free-Response Questions

Complete this exercise to build and practice the skills you will need to answer free-response questions on the exam. Use a separate sheet of paper if necessary.

1. Air pollutants come from a variety of sources.

   (a) Identify a piece of legislation that regulates criteria air pollutants.

   (b) Identify the six criteria air pollutants regulated in the legislation you named in (a).

2. The Great Smoky Mountains were named for the natural air pollutants that reduce visibility and give the landscape a smoky appearance.

   (a) Identify a primary pollutant that is responsible for photochemical smog formation.

   (b) Explain how photochemical smog is formed due to primary pollutants.

3. Radon, mold, carbon monoxide, and particulate matter are common indoor air pollutants.

   (a) For radon identify the source of the indoor air pollutant, and one way that an individual may reduce their possible exposure.

   (b) For carbon monoxide, identify the source of the indoor air pollutant, and one way that an individual may reduce their possible exposure.

4. Acid deposition can have impacts on ecosystems as well as on man-made structures.

    (a) Identify the primary pollutants involved in acid deposition.

    (b) Explain how these primary pollutants enter the atmosphere.

    (c) Describe how acid deposition forms as a result of these primary pollutants.

    (d) Explain how secondary pollutants can negatively affect amphibians.

5. Noise pollution occurs in many places.

    (a) Describe the difference in noise pollution in wealthier parts of cities and lower-income areas.

    (b) Describe noise pollution in suburban or rural areas.

Unit 7 Multiple-Choice Review Exam

1. Household bleach with a pH of 13 is how many times more basic than pure water with a pH of 7?
    (a) 100
    (b) 1,000
    (c) 6,000
    (d) 1,000,000

2. Which of the following shows the correct formation of a secondary pollutant from a primary pollutant?
    (a) VOCs + sunlight → ozone
    (b) $NO_2 + H_2O → HNO_3$
    (c) $H_2O + CO_2 → H_2CO_3$
    (d) $O_3 → O^- + O_2$

3. The combustion of coal results in the release of which of the following air pollutants?
    (a) ozone
    (b) CFCs
    (c) sulfur dioxide
    (d) H_2O

4. Which of the following describes an ecosystem service that has been degraded from large amounts of acid deposition?
    (a) Declines in forests due to negative impacts on the red spruce tree could result in a decrease in carbon sequestration.
    (b) Increase in ocean acidification results in a decline in fisheries.
    (c) Amphibians may be unable to tolerate the changes in pH and may fail to reproduce.
    (d) Metals such as aluminum and mercury may cause an increase in crop yields.

5. In a coal burning power plant, an electrostatic precipitator is designed to remove
    (a) sulfur dioxide.
    (b) nitrogen dioxide
    (c) particulate matter.
    (d) carbon dioxide.

6. A thermal inversion causes severe pollution events. Thermal inversions
    (a) occur when a warm inversion layer traps emissions that then accumulate beneath it.
    (b) occur when a cold inversion layer traps emissions that then accumulate beneath it.
    (c) occur when a warm inversion layer traps emissions that then accumulate above it.
    (d) occur when cold inversion layer traps emissions that then accumulate above it.

7. Which location would most likely experience photochemical smog?
    (a) Oslo, Norway
    (b) Los Angeles, California
    (c) Manaus, Brazil
    (d) Vancouver, British Columbia

8. Which of the following correctly matches the criteria air pollutant to its human-derived source?
   (a) particulate matter: secondary pollutant formed by combination of light, oxygen, and VOCs
   (b) lead: combustion of fuels containing sulfur
   (c) nitrogen oxides: additive of old paint
   (d) CO: incomplete combustion of fossil fuels including malfunctioning exhaust systems

9. Why might city planners attempt to limit the number of bakeries, dry cleaners, and power plants in a small area?
   (a) These activities release lead which may get into the water system and bioaccumulate in the food chain.
   (b) These activities release VOCs which may serve as a precursor to tropospheric ozone formation.
   (c) These activities may release particulate matter that can contribute to haze and smog.
   (d) These activities release mercury which can negatively impact the central nervous system.

10. Which of the following represents why children may be more susceptible to indoor air pollution and carbon monoxide poisoning?
    (a) They have less developed respiratory systems.
    (b) They have less developed nervous systems.
    (c) They have a higher respiration rate.
    (d) They have a lower respiration rate.

11. Which of the following represents the best solution to dealing with secondary air pollutants?
    (a) Reduce the production of primary pollutants.
    (b) Encourage families to move to areas where secondary pollutants aren't found.
    (c) Increase the production of primary pollutants using scrubbers or catalytic converters.
    (d) Remove natural sources of primary pollutants such as volcanoes.

Use the following figure to answer question 12.

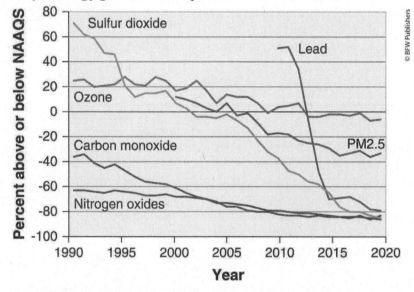

12. Which of the following represents the correct way to calculate the percent decrease in sulfur dioxide emissions from 1990 to 2019?
    (a) $(-78 - 70) \div 70 \times 100$
    (b) $70 - (-78) \div (-78) \times 100$
    (c) $70 \div (-78 - 70) \times 100$
    (d) $(-78 - 70) \div 70 \times 10$

13. Which piece of legislation regulates the average ozone concentration over an eight-hour period?
    (a) State of the Air
    (b) Lacey Act
    (c) Clean Air Act
    (d) Paris Agreement

14. Which of the following accurately describes the trend in nitrogen oxides throughout the day?
    (a) Nitrogen oxides start high and then peak in the afternoon before falling again as individuals drive less in the mornings than the evenings.
    (b) Nitrogen oxides rise steadily throughout the morning and fall in the afternoon due to the behavior of the atmosphere.
    (c) Ozone levels are typically lowest in the mornings and the late evenings as transportation rates are the lowest.
    (d) There is time delay between the release of nitrogen oxides and the formation of ozone.

15. Nitrogen oxides are most likely to peak in the mornings and again in the evenings because:
    (a) Ozone requires the release of NOx to form.
    (b) Sunlight is required to catalyze the reaction of NOx into NO and O^-.
    (c) Ozone will eventually breakdown throughout the day.
    (d) NOx is released by cars during the morning and afternoon commutes

16. Photochemical smog is mostly due to:
    (a) the combination of sulfate compounds, particulate matter and fog.
    (b) the combination of NO_x and SO_x with water vapor.
    (c) the combination of H_2O with carbon dioxide from cars.
    (d) the reaction between ozone and photochemical oxidants.

17. Why is temperature a factor in the formation of smog?
    (a) It slows down the reaction of SO_x with water vapor to create sulfuric acid.
    (b) VOC evaporation increases as temperatures increase.
    (c) Higher temperature days cause less electricity to be produced, which releases NO_x.
    (d) Longer days result in a decrease in light which drives up the formation of ozone.

18. Which of the following is a human health problem associated with photochemical oxidants?
    (a) Dizziness due to low oxygen levels.
    (b) Difficulty taking deep breaths.
    (c) Decline in photosynthesis rates.
    (d) Decreased neurological functioning

19. Particulate matter may be difficult to control because:
    (a) it has both an anthropogenic and natural source.
    (b) it is derived from activities such as dry cleaning.
    (c) it causes a decline in photosynthesis rates.
    (d) it causes an overall increase in crop yields which is beneficial for farmers.

20. Radon detectors are not mandated in all homes in the United States due to differences in radon concentrations around the country. Which of the following describes the source of Radon?
    (a) Radon comes from oil processing which is a major industry in the Gulf of Mexico.
    (b) Radon is naturally occurring in all large bodies of water.
    (c) Radon naturally occurs in the bedrock of all parts of the United States.
    (d) Radon is an anthropogenic secondary pollutant caused by bakeries.

21. Which of the following represents a human health concern with living in buildings whose walls were painted before the 1960s?
    (a) Mercury could be released which bioaccumulates and biomagnifies in the food chain.
    (b) Nitrogen oxides are released as fumes from older paints and is tied to eye irritation.
    (c) Lead is found in old paint which may cause brain damage if chips are consumed.
    (d) SO_2 may be released which can cause chronic lung conditions.

Use the following figure to answer questions 22-24.

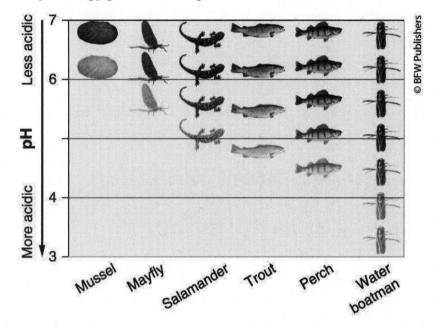

22. Water boatman may be found in a variety of ecosystems suffering from acid deposition because:
    (a) They are only able to live in less acidic conditions.
    (b) They are able to outlive perch.
    (c) They have an intolerance to more acidic conditions.
    (d) They have a wide range of tolerance for pH levels.

23. Which of the following aquatic organisms would be able to survive but not reproduce at a pH of 4.5?
    (a) water boatman
    (b) perch
    (c) mayfly
    (d) mussel

24. Given the data provided what is the threshold for all the aquatic organisms to survive and reproduce?
    (a) 3
    (b) 4
    (c) 5
    (d) 7

25. Which of the following represents an environmental problem associated with the use of long-range sonar systems?
    (a) Terrestrial animals may be unable to hear mating calls.
    (b) Aquatic organisms may have increased reproduction rates.
    (c) Whales may experience beaching events.
    (d) Amphibians are unable to tolerate the noise and may see decreased reproduction,

26. Decreased sleep quality and quantity as a result of noise pollution can result in which of the following long-term outcomes for human health?
    (a) high blood pressure
    (b) auditory cancer
    (c) asthma
    (d) sick building syndrome

27. Vapor recovery nozzles aim to
    (a) reduce CO_2 emissions.
    (b) reduce VOC emissions.
    (c) reduce SO_2 emissions.
    (d) reduce particulate matter.

28. Prior to the 2008 Beijing Olympics, the city increased public transportation in an effort to
    (a) reduce the amount of CO_2 released preventing acid deposition.
    (b) reduce the amount of SO_2 released preventing ocean acidification.
    (c) reduce the amount of ozone produced as a byproduct of coal combustion.
    (d) reduce the amount of NO_2 released that is a precursor to ozone formation.

29. Cap-and-trade amendments to the Clean Air Act have successfully reduced SO_2 emissions in some areas. However, a disadvantage associated with this solution is that
    (a) companies can increase their annual emissions as long as they pay to do so.
    (b) individuals living in wealthier cities are subjected to increased amounts of sulfur emissions.
    (c) hazardous wastes from nearby landfills are more likely to be found in lower income areas.
    (d) the benefits of the amendments are more likely to be seen in lower income areas.

30. Catalytic converters utilize a honeycomb device cotated in metal that is damaged by lead. Which of the following best describes how this was addressed by the Clean Air Act?
    (a) A cap-and-trade system was implemented for sulfur dioxide.
    (b) Lead was no longer allowed to be an additive to gasoline.
    (c) Electrostatic precipitators were added to prevent the release of particulate matter.
    (d) Carbon monoxide detectors were added to all public buildings and large apartment buildings.

UNIT 8

Aquatic and Terrestrial Pollution

Unit Summary

Pollution in aquatic and terrestrial ecosystems is generally defined as the contamination of land, water bodies, or groundwater with substances produced through human activities. In this unit you will learn to recognize various forms of pollutants and the negative effects of pollutants on ecosystems and food webs as well as the health impacts on humans. You will identify remediation plans for waste and contamination. Being able to identify the sources and types of pollutants, allows us to stop, slow, or remediate the negative consequences of pollutants in our environment.

MODULES IN THIS UNIT

Module 47: Sources of Pollution, Human Impacts on Ecosystems, and Endocrine Disruptors
Module 48: Human Impacts on Wetlands and Mangroves, Eutrophication, and Thermal Pollution
Module 49: Persistent Organic Pollutants (POPs), Bioaccumulation, and Biomagnification
Module 50: Solid Waste Disposal
Module 51: Waste Reduction Methods
Module 52: Sewage Treatment
Module 53: Lethal Dose 50% (LD_{50}) and Dose Response Curves
Module 54: Pollution, Human Health, Pathogens, and Infectious Diseases

---

Unit Opening Case: *Paper, Plastic or Reusable?*
The case study reinforces the idea that there are no easy answers when it comes to solid waste and recycling. Although you may be familiar with some of the concepts in this chapter and have ideas about the benefits of recycling, this chapter shows you how to approach the issues of solid waste in a scientific way.

Do the Math
This unit contains the following "Do the Math" boxes to help prepare you for calculation questions you might encounter on the exam.
- "Calculating the Magnitude of Oil Pollution" (page 561)
- "Estimating Half-Lives of Toxic Chemicals" (page 577)
- "How Much Leachate Might Be Collected?" (page 590)
- "Calculating Recycling Rates" (page 602)
- "Building a Manure Lagoon" (page 615)
- "Estimating LD_{50} Values and Safe Exposures" (page 621)

To make sure you understand the concepts and techniques presented in these boxes, do the practice problems presented in the text as well as the additional "Practice the Math" problems that appear in Module 47, Module 49, Module 50, Module 51, Module 52, and Module 53 of this study guide.

MODULE 47: Sources of Pollution, Human Impacts on Ecosystems, and Endocrine Disruptors

Before You Read the Module

Focus on Learning Goals

Use the module learning goals to guide your reading. On a separate piece of paper, write down each goal and take notes to help you meet each learning goal. After studying this module, you should be able to:

- 47-1 identify point and non-point sources of pollution.
- 47-2 explain why species differ in their tolerance to pollutants.
- 47-3 identify the major groups of chemical pollutants and where they come from.
- 47-4 describe the impacts of chemical pollutants.
- 47-5 identify the sources of oil pollution.

Key Terms

Point source	Neurotoxin	Endocrine disruptor
Nonpoint source	Carcinogen	Wastewater
Homeostasis	Mutagen	
Polychlorinated biphenyls	Teratogen	
(PCBs)	Allergen	

While You Read the Module

Answer the following questions as you read. Use a separate sheet of paper if necessary.

Unit Opening Case: Paper, Plastic, or Reusable?

1. Pre-reading question: Do you think it would be better for the environment to drink a hot beverage from a paper cup, Styrofoam cup, or a reusable cup? Explain why.

2. Describe polystyrene.

3. List the benefits of using polystyrene.

4. List the disadvantages of using polystyrene cups.

5. How did Dunkin' Donuts assist in reducing the use of Styrofoam containers?

6. What is a disadvantage of using paper cups instead of Styrofoam cups?

7. Identify the factors used to consider the environmental cost of a type of cup.

8. List the inputs and outputs of the manufacture and use of a paper cup compared to a polystyrene cup.

9. List the inputs and outputs of the manufacture and use of a polystyrene cup compared to a paper cup.

10. What factors did researchers examine in 2021 to study the impact of single-use and multiple-use items?

11. List the findings of the impact research.

12. After reading the case study, what do you think is the best choice of cups for hot beverages? Explain your reasoning.

Module 47: Sources of Pollution, Human Impacts on Ecosystems, and Endocrine Disruptors

13. Describe how pollution in aquatic and terrestrial ecosystems is defined.

Point sources of pollution have single locations while nonpoint sources have diffuse locations

14. Define point source and give an example.

15. Define nonpoint source and give an example.

16. Figure 47.1(b): Identify two possible nonpoint source pollutants from the surrounding areas into the water system.

17. Explain why a distinction between point source and nonpoint source pollution is important.

Organisms differ in their tolerance to various pollutants

18. Define homeostasis and explain why it is necessary for an organism.

19. Describe the possible effects on an animal of exposure to low, medium, and high concentrations of a pesticide.

20. Explain the possible meaning of the absence of mayflies from a stream where they usually are found.

21. List the various pollutants and human activities that affect coral reef biodiversity.

Chemical pollutants include heavy metals and synthetic compounds that are produced naturally and by human activities

Heavy Metals

22. Describe how lead can become a contaminant in drinking water.

23. Identify who is most sensitive to lead and its detrimental effects.

24. Describe the federal guidelines about lead implemented in the last three decades.

25. Describe the series of steps leading up to the Flint, Michigan lead contamination in 2014.

26. Identify the EPA guidelines for lead in water.

27. Identify the highest concentration of lead found in the Flint water.

28. How did Flint try to remedy the contamination situation in 2015?

29. Identify how many children have been affected by the contaminated water in Flint.

30. How much did Flint, Michigan have to pay to clean up the contamination?

31. Describe how arsenic can enter groundwater naturally and through anthropogenic activities.

32. Identify how arsenic is removed from drinking water.

33. Figure 47.3: Identify the part of the United States that has the lowest concentrations of arsenic and state the levels of concentration.

34. Identify the detrimental effects of arsenic on humans.

35. Identify and describe the safe limits set for arsenic in 1999.

36. Describe how arsenic contaminated the water in Bangladesh and India.

37. Explain the possible remedies for the contaminated water in Bangladesh and India.

38. Figure 47.4: List the regions by the amount of mercury emissions from greatest to least and name the percentages.

39. List human activities associated with the release of mercury.

40. Describe the reason for mercury and lead pollution from petroleum exploration.

41. Describe how methylmercury develops from inorganic mercury.

42. Identify the detrimental effects of methylmercury on humans.

43. Explain how humans are exposed to methylmercury.

44. Identify the United States Environmental Protection Agency limit for methylmercury per gram of tuna.

45. Figure 47.5: Identify the location of the yellowfin tuna with the highest levels of methylmercury.

46. Identify the industries where mercury pollution could be reduced.

Synthetic Organic Compounds

47. Identify the benefits of pesticides.

48. Identify when synthetic pesticides were first developed.

49. Describe the use of the insecticide endosulfan and the unintended consequences of its use.

50. Describe pesticide inert ingredients and give an example.

51. Explain how the herbicide Roundup works and its unintended consequences of its use.

52. Explain why people did not know about this side effect of Roundup until recently.

53. Figure 47.7: List the contaminants found in streams from highest to lowest frequency.

54. Describe perchlorates.

55. Describe how humans are exposed to perchlorates and the possible negative effects.

56. Describe the detrimental effects from dumping industrial compounds into the Cuyahoga River of Ohio.

57. Define polychlorinated biphenyls (PCBs).

58. Explain why PCBs are dangerous.

59. Describe PBDEs (polybrominated diphenyl ethers).

60. Identify where PBDEs are being detected and their possible detrimental effects.

61. Describe PFAS (per- and polyfluoroalkyl substances).

62. How are PFAS a public concern?

Chemical pollutant impacts can be categorized as neurotoxins, carcinogens, teratogens, allergens, or endocrine disruptors

63. Table 47.1: Copy the table.

Neurotoxins

64. Define neurotoxin.

65. Explain how a pesticide neurotoxin affects insects and other invertebrates.

Carcinogens

66. Define carcinogen.

67. Define mutagen.

68. List some of the most well-known carcinogens.

Teratogens

69. Define teratogen.

70. Identify two sources of teratogens.

71. Describe fetal alcohol syndrome.

Allergens

72. Define allergen.

73. Identify common chemicals that cause allergic reactions.

Endocrine Disruptors

74. Define endocrine disruptor.

75. Figure 47.9: Explain how hormone disrupting chemicals cause a negative effect on an organism.

76. Define wastewater.

77. Explain the possible effects of reproductive hormones on animals found in waterways.

78. Describe how the fish in the Chesapeake Bay were affected by the sewage contamination.

79. Identify possible concerns for humans regarding the reproductive hormones in the waterways and other endocrine disruptors.

Oil pollution comes from multiple sources

80. Identify three environmental problems created by oil pollution.

Sources of Oil Pollution

81. Describe the *Exxon Valdez* tanker accident that led to an oil spill.

82. List the detrimental environmental effects from the *Exxon Valdez* oil spill.

83. Describe the effects from the *Exxon Valdez* still seen twenty years later.

84. Explain the new North American tanker rules.

85. Identify the approximate number of offshore drilling platforms located around the world.

86. Describe how the 2010 BP oil leak started.

87. Identify how much oil was lost. How long did it take to seal the leak?

88. Identify the difference between the *Exxon Valdez* and BP oil spills.

89. Describe an oil seep.

90. Figure 47.11(b): List the sources of oil pollution in the world's oceans from greatest to least with percentages.

Remediating Oil Pollution

91. List and describe the four ways to remediate oil pollution.

92. Identify and describe detrimental effects of oil remediation methods.

93. Identify new methods being researched for oil spill clean ups.

94. Describe the moving oil plume from the BP explosion and the method of removal.

95. Although cleaning oil-contaminated shorelines is better than not cleaning them, there are some negative effects. Describe the negative effects of cleaning shorelines with high pressure hot water.

Practice the Math: Calculating the Magnitude of Oil Pollution

Read "Do the Math: Calculating the Magnitude of Oil Pollution" on page 561. Try "Your Turn." For more math practice, do the following problems. Remember to show your work. Use a separate sheet of paper if necessary.

In 1989, the *Exxon Valdez* oil tanker crashed into a reef and spilled approximately 34 thousand metric tons of crude oil into Alaskan waters. On the low estimate, each year approximately three hundred thousand metric tons of oil enter Alaskan waters through natural seeps.

(a) Convert the tons of oil to gallons

     i.   Spilled into Alaskan waters.

     ii.   Seeped into Alaskan waters.

(b) An average household in Alaska may use 1,200 gallons of heating oil per year. How many Alaskan homes could be heated by the gallons of oil that spilled and seeped into Alaskan waters?

     i.   Spilled into Alaskan waters.

     ii.   Seeped into Alaskan waters.

Review Key Terms

Match the key terms on the left with the definitions on the right.

_____ 1. Point source

_____ 2. Nonpoint source

_____ 3. Homeostasis

_____ 4. Polychlorinated biphenyls (PCBs)

_____ 5. Neurotoxin

_____ 6. Carcinogen

_____ 7. Mutagen

_____ 8. Teratogen

_____ 9. Allergen

_____ 10. Endocrine disruptor

_____ 11. Wastewater

a. A group of industrial compounds that were once used to manufacture plastics and insulate electrical transformers.

b. A chemical that interferes with the normal development of embryos or fetuses.

c. A chemical that causes allergic reactions.

d. A diffuse area that produced pollution.

e. The water produced by livestock operations and human activities, including human sewage from toilets and gray water from bathing and washing clothes and dishes.

f. A chemical that disrupts the nervous systems of animals.

g. A chemical that interferes with the normal functioning of hormones in an animal's body.

h. The ability to experience relatively stable internal conditions in their bodies.

i. A type of carcinogen that causes damage to the genetic material of a cell.

j. A distinct location from which pollution is directly produced.

k. A chemical that causes cancer.

MODULE 48: Human Impacts on Wetlands and Mangroves, Eutrophication, and Thermal Pollution

Before You Read the Module

Focus on Learning Goals

Use the module learning goals to guide your reading. On a separate piece of paper, write down each goal and take notes to help you meet each learning goal. After studying this module, you should be able to:

- 48-1 describe the human impacts of water availability.
- 48-2 explain the human impacts on wetlands and mangroves.
- 48-3 explain the causes and consequences of eutrophication and sediments.
- 48-4 describe the impacts of thermal and noise pollution.

Key Terms

Levee	Aqueduct	Oxygen sag curve
Dikes	Desalination (Desalinization)	Thermal pollution
Dam	Distillation	Thermal shock
Reservoir	Reverse osmosis	
Fish ladder	Eutrophication	

While You Read the Module

Answer the following questions as you read. Use a separate sheet of paper if necessary.

Module 48: Human Impacts on Wetlands and Mangroves, Eutrophication, and Thermal Pollution
Humans are altering the availability of water by controlling its movement

Levees and Dikes

1. Define levee and describe its purpose.

2. List how levees have created environmental challenges.

3. Describe the events of the levee failure in New Orleans in 2005.

4. Define dike.

5. Identify where dikes are most common.

6. Describe how people in the Netherlands successfully live and farm in areas that should be under water.

Dams

7. Define dam.

8. Define reservoir.

9. Identify the largest reservoir system in the United States.

10. Explain how humans have used dams for our benefit.

11. Describe the development of the Three Gorges Dam.

12. List the benefits of the Three Gorges Dam.

13. List the costs of the Three Gorges Dam to people and the environment.

14. List the environmental problems associated with dams.

15. Define fish ladder.

16. Describe how the Klamath River has changed over the past 100 years.

17. Explain the decision made in 2009 about the Klamath River dams.

18. When is the final dam removal expected to occur?

Aqueducts

19. Define aqueduct.

20. Describe how long aqueducts have been used.

21. Describe the aqueduct systems of New York City and Los Angeles.

22. Figure 48.5: Analyze the photograph and identify a negative effect of the open aqueduct.

23. Identify the negative impacts of an aqueduct system.

24. Identity the benefits of an aqueduct system.

25. Describe the impacts of the water diversion projects between India and Bangladesh.

26. Describe how China's water diversion projects could impact India and Bangladesh.

27. Explain the Soviet Union water diversion project and the outcome of the Aral Sea in Central Asia.

28. List the environmental impacts the water diversion project had on the Aral Sea.

29. What was the status of the South Aral Sea as of 2009?

Humans are converting salt water into fresh water by desalination

30. Define desalination (desalinization).

31. Identify the importance of desalination in the Middle East.

32. Define distillation.

33. Identify the disadvantages of distillation.

34. Define reverse osmosis.

35. Describe brine and the environmental consequence of brine.

36. Figure 48.8: Identify the areas with the lowest amounts of available fresh water.

37. Identify how many people live in areas with water scarcity.

Humans are impacting wetlands and mangroves through development, dams, overfishing, and pollutants

38. List the ecosystem services provided by wetlands.

39. Explain why freshwater wetlands have been or continue to be drained.

40. Describe the effect of commercial fishing on the wetlands of the Amazon and Pantanal in South America.

41. According to the United Nations, how fast are the world's wetlands disappearing?

Aquatic ecosystems are being harmed by eutrophication and sediment inputs

42. Define eutrophication.

Algal Blooms

43. Identify the two most important nutrients in an aquatic ecosystem.

44. Describe an algal bloom and how it develops.

45. Explain the harmful effects of an algal bloom.

46. Describe water that is hypoxic.

47. Describe how dead zones on the coasts have changed in the last century.

48. Figure 48.11: Describe the region in the United States with the most dead zones.

49. Describe how eutrophic lakes become hypoxic.

Oxygen Sag Curves

50. Define oxygen sag curve.

51. Figure 48.12: Draw the diagram. Label all parts of the diagram, including the x and y axis labels.

52. How does the detection of oxygen sag curves help researchers?

Sediments

53. Explain where the sediments in streams and rivers originate.

54. Describe the effects of increased sediment in the waterways.

Thermal pollution comes from warming water bodies while noise pollution comes from human sounds

    55. Define thermal pollution.

Thermal Pollution

    56. Describe how logging in forests can cause warming of streams, rivers, and wetlands.

    57. Identify the types of industries involved with using water as a cooling agent.

    58. Define thermal shock.

    59. Describe how thermal shock could lead to the death of an animal.

    60. Explain a possible solution to stop thermal pollution.

    61. Figure 48.14: List the two cooling methods used at the Three Mile Island nuclear plant.

Noise Pollution

    62. Describe noise pollution in the ocean.

    63. How have the U.S. Navy's operations caused concern for ocean life?

    64. Identify how some ship builders are helping with noise pollution.

Review Key Terms

Match the key terms on the left with the definitions on the right.

_____ 1. Levee

a. A canal, ditch, or pipe used to carry water from one location to another.

_____ 2. Dikes

b. Structures built to prevent ocean waters from flooding adjacent land.

_____ 3. Dam

c. A process of desalination in which water is boiled and the resulting steam is captured and condensed to yield pure water.

_____ 4. Reservoir

d. Excess nutrients from human activities that make their way into waterbodies; it causes nutrient pollution that alters food webs and harms water quality.

_____ 5. Fish ladder

e. An enlarged bank built up on each side of the river.

_____ 6. Aqueduct

f. A dramatic change in temperature that can kill many species.

_____ 7. Desalination (Desalinization)

g. A process for obtaining fresh water by removing the salt from salt water.

_____ 8. Distillation

h. Occurs when humans cause a substantial change in the temperature of a water body.

_____ 9. Reverse osmosis

i. A stair-like structure with water flowing over them, which allows migrating fish to get around a dam.

_____ 10. Eutrophication

j. A process of desalination in which water is forced through a thin semipermeable membrane at high pressure.

_____ 11. Oxygen sag curve

k. A barrier that runs across a river or stream to control the flow of water.

_____ 12. Thermal pollution

l. The relationship of oxygen concentrations to the distance from a point source of decomposing sewage or other pollutants.

_____ 13. Thermal shock

m. The water body created by damming a river or stream.

MODULE 49: Persistent Organic Pollutants (POPs), Bioaccumulation, and Biomagnification

Before You Read the Module

Focus on Learning Goals

Use the module learning goals to guide your reading. On a separate piece of paper, write down each goal and take notes to help you meet each learning goal. After studying this module, you should be able to:

- 49-1 describe how persistent organic pollutants affect ecosystems.
- 49-2 explain how routes of exposure and solubility determine the concentrations of chemicals that organisms experience.
- 49-3 explain how bioaccumulation and biomagnification can increase the concentrations of chemicals in organisms.

Key Terms

Persistence	Route of exposure	Biomagnification
Persistent organic pollutants (POPs)	Solubility	
	Bioaccumulation	

While You Read the Module

Answer the following questions as you read. Use a separate sheet of paper if necessary.

Module 49: Persistent Organic Pollutants (POPs), Bioaccumulation, and Biomagnification

Some chemicals can persist in nature and harm ecosystems for years

Chemical Persistence

1. Define persistence.

2. Identify the factors that can affect persistence.

3. Table 49.1: What is the half-life in the environment of DDT compared to malathion?

4. After spraying DDT 60 years ago, how much of the chemical could be present in the soil today?

5. Define persistent organic pollutants (POPs).

Persistent Organic Chemicals of High Concern

6. When did the manufacturing of PCBs stop in the United States? Explain why they are still a concern.

7. Describe the environmental damage caused by General Electric in New York State and how the company had to remedy the situation.

8. Identify the negative health effects of PFAS.

9. Explain why PFAS are known as forever chemicals.

10. List the two solutions when groundwater has been contaminated with PFAS.

Concentrations of chemicals experienced by organisms depends on routes of exposure and solubility

Routes of Exposure

11. Define route of exposure.

12. Figure 49.2: List the different routes of exposure from the mother to the fetus/baby.

13. Explain bisphenol A and describe its route of exposure in children.

14. Identify the possible effects of bisphenol A in children.

15. List routes of exposure for animals.

Solubility of Chemicals

16. Define solubility.

17. Describe the environmental difference between water soluble and fat or oil soluble substances.

Bioaccumulation and biomagnification can dramatically increase the concentration of a chemical in organisms by storing them in fat

18. Define bioaccumulation.

19. Describe bioaccumulation of methylmercury in fish.

20. Figure 49.3: Explain how over time fish accumulate a higher concentration of a chemical.

21. Define biomagnification.

22. Figure 49.6: List the components of the food chain and their DDT values.

23. How many times higher is the DDT concentration in the osprey than the water?

24. Identify the negative consequences of DDT on fish-eating birds.

25. Describe the reason fish-consumption advisories are set.

Practice the Math: Estimating Half-Lives of Toxic Chemicals

Read "Do the Math: Estimating Half-Lives of Toxic Chemicals" on page 630. Try "Your Turn." For more math practice, do the following problem. Remember to show your work. Use a separate sheet of paper if necessary.

Capsaicin, the chemical that makes chili peppers "hot," is often used as a pesticide. Dicambia is a broad-spectrum herbicide that can be applied to the leaves of weeds or soil. Methoprene is an insecticide that disrupts juvenile insect growth. Each of these pesticides has a range of potential half-life length based on environmental conditions.

Substance	Half-Life Range
Capsaicin	2–8 days
Dicamba	30–60 days
Methoprene	10–14 days

Data from http://npic.orst.edu/factsheets/half-life.html

Calculate the number of days it would take to have 1/16 of the original pesticide in the most persistent and least persistent conditions.

(a) How many half-lives would occur?

(b) Determine the number of days for each of the following.

    i.   Capsacin high persistence: 8 days

    ii.   Capsacin low persistence: 2 days

    iii. Dicamba high persistence: 60 days

    iv.   Dicamba low persistence: 30 days

    v.   Methoprene high persistence: 14 days

    vi.   Methoprene low persistence: 10 days

After You Read the Module

Review Key Terms
Match the key terms on the left with the definitions on the right.

_____ 1. Persistence

a. The length of time a chemical remains in the environment.

_____ 2. Persistent organic pollutants (POPs)

b. The way in which an individual might come into contact with an environmental hazard, such as a chemical.

_____ 3. Route of exposure

c. Synthetic, carbon-based molecules that break down very slowly in the environment.

_____ 4. Solubility

d. The selective absorption and concentration of a chemical within an organism over time.

_____ 5. Bioaccumulation

e. The increase in chemical concentration in animal tissues as the chemical moves up the food chain.

_____ 6. Biomagnification

f. How well a chemical dissolves in a liquid.

MODULE 50: Solid Waste Disposal

Before You Read the Module

Focus on Learning Goals

Use the module learning goals to guide your reading. On a separate piece of paper, write down each goal and take notes to help you meet each learning goal. After studying this module, you should be able to:

- 50-1 describe how solid waste production has changed over time and identify the sources.
- 50-2 explain how landfills are used to dispose of municipal solid waste.
- 50-3 describe how incineration is used to dispose of municipal solid waste.
- 50-4 explain why some municipal solid waste does not go to landfills or incinerators.
- 50-5 identify how hazardous waste is safely disposed.

Key Terms

Solid waste
Municipal solid waste
(MSW)
Waste stream
Leachate

Sanitary landfill
Tipping fee
Incineration
Ash
Waste-to-energy

Hazardous waste
Superfund Act
Brownfields

While You Read the Module

Answer the following questions as you read. Use a separate sheet of paper if necessary.

Module 50: Solid Waste Disposal

1. Define solid waste.

Solid waste pollution has increased over time from residences, businesses, industries, and agricultural activities

2. Explain the lifecycle of a broken bookcase from 1900.

3. Describe what happened with consumption patterns in the United States after World War II.

Municipal Solid Waste

4. Define municipal solid waste (MSW).

5. List the two main sources to MSW and the percentage each contributes.

6. Figure 50.1: Describe the per capita trend of municipal solid waste in the United States from 1960 to 2020.

7. Compare average waste generation per person per day in the United States with the rest of the developed world.

8. Figure 50.2: Identify how many individuals are in the photo and describe why the individuals are in the dump.

9. Explain why developing countries have become responsible for a greater portion of global municipal solid waste.

10. Define waste stream.

11. Identify what materials are included in "paper and paperboard."

12. Figure 50.3: List each category of MSW and its percentage from greatest to least.

Electronic Waste

13. List the items considered to be electronic waste.

14. Identify the amount of metals contained in an older-style cathode-ray-tube television or computer monitor.

15. Why don't more people in the United States recycle electronics?

16. Describe what happens to e-waste in the United States after it has been sent to be recycled.

Landfills are the primary destination for municipal solid waste

17. Figure 50.5: Identify the percentage of municipal solid waste that ends up in landfills and the percentage that is incinerated.

A Brief History of Landfills

18. Describe an open landfill.

19. Describe the negative consequences of open landfills.

Landfill Design

20. Describe two processes that occur in landfills.

21. Define leachate.

21. Define sanitary landfill.

22. Describe how a sanitary landfill is constructed.

23. Figure 50.6: What happens to methane produced in a sanitary landfill.

24. Explain why water entering the landfill should be kept to a minimum.

25. Explain how leachate is monitored.

26. List items that are acceptable to go into a landfill.

27. List items that should not go into a landfill.

28. Explain how the volume of MSW added to a landfill is reduced.

29. Identify how a sanitary landfill can be reclaimed.

30. Figure 50.7: Explain the life and reclamation of the Fresh Kills landfill in Staten Island, New York.

31. Define tipping fee.

32. Identify the average cost of tipping fees for sanitary landfills.

33. Explain the consequences if tipping fees are too high.

Choosing a Landfill Site

34. Identify the requirements for siting a landfill.

35. Identify the meaning of NIMBY. Explain the significance of NIMBY in regards to the siting of a landfill.

36. Discuss the issues of environmental injustice with the Adams Center Landfill in Fort Wayne, Indiana. Describe the resolution.

Environmental Consequences of Landfills

37. List the environmental consequences of landfills.

Incinerators reduce solid waste by burning it

38. Define incineration.

39. Identify the amount by which an incinerator can reduce the volume and weight of solid waste.

Incineration Basics

40. Figure 50.8: List the steps for incineration of waste.

41. Define ash.

42. Describe the disposal or uses of non-toxic and toxic ash.

43. Explain how exhaust gases are collected and disposed of.

44. Define waste-to-energy.

Environmental Consequences of Incineration

45. Describe the economic considerations of incineration.

46. Explain why incinerators may not completely burn all waste.

A lot of solid waste ends up in the ocean

47. Explain why solid waste is not always properly disposed.

48. Figure 50.9: Identify the two garbage patches in the Pacific Ocean.

49. What event caused the United States to end ocean dumping?

50. Figure 50.10: Describe why some countries are not able to prevent dumping garbage into the ocean or rivers.

Hazardous waste requires proper handling and disposal

Producing Hazardous Wastes

51. Define hazardous waste.

52. List and describe the four characteristics of hazardous waste.

53. Differentiate between hazardous waste and toxic waste.

54. List what sources generate hazardous waste.

Disposing of Hazardous Waste

55. Describe what must happen to hazardous waste in order to carry out proper disposal.

56. Describe how disposal of a $5 can of oven cleaner might eventually cost $25 to $50. Who is responsible for the higher costs?

57. Describe the composition of coal ash and coal tailings and how disposal of these wastes are treated.

Regulation and Oversight of Handling Hazardous Waste

58. Identify the two federal acts that cover the handling of hazardous waste.

59. Explain the Resource Conservation and Recovery Act (RCRA).

60. Explain the Hazardous and Solid Waste Amendments (HSWA).

61. Define Superfund Act.

62. Figure 50.2: Identify your state of residence and the density of proposed and finalized Superfund sites.

63. How many Superfund sites had been remediated as of 2021?

64. Explain what led to Love Canal in Niagara Falls, New York becoming a Superfund site.

65. List and describe the timeline of events for the Love Canal Superfund site.

66. Identify the problem that has prevented implementation of CERCLA.

Brownfields

67. Define Brownfields.

68. Explain how Brownfields are different from Superfund sites. Give an example.

69. Identify the reason the Brownfields Program has been criticized for being inadequate.

Practice the Math: How Much Leachate Might Be Collected?

Read "Do the Math: How Much Leachate Might Be Collected?" on page 590. Try "Your Turn." For more math practice, do the following problems. Use a separate sheet of paper if necessary.

The annual precipitation at a landfill is 250 mm per year, and 50 percent of this water runs off the landfill.

(a) If the landfill has a surface area of 10,000 m², calculate the volume of water in cubic meters that infiltrates the landfill per year.

(b) How much volume of leachate in m³ is treated per year if the leachate collection system is 80 percent effective?

After You Read the Module

Review Key Terms
Match the key terms on the left with the definitions on the right.

_____ 1. Solid waste

    a. The process of burning waste materials to reduce volume and mass, and sometimes to generate electricity or heat.

_____ 2. Municipal solid waste (MSW)

    b. Liquid that can contain elevated levels of pollutants as a result of having passed through the solid waste of a landfill.

_____ 3. Waste stream

    c. The waste produced by humans as discarded materials that is not in liquid or gas form and do not pose a toxic hazard to humans and other organisms.

_____ 4. Leachate

    d. Liquid, solid, gaseous, or sludge waste material that is harmful to humans, ecosystems, or materials.

_____ 5. Sanitary landfill

    e. A system in which heat generated by incineration is used as an energy source rather than released into the surrounding environment.

_____ 6. Tipping fee

    f. A fee charged for trucks that deliver and tip solid waste into a landfill or incinerator.

_____ 7. Incineration

    g. The common name for the Comprehensive Environmental Response, Compensation, and Liability Act (CERCLA); a 1980 U.S. federal act that imposes a tax on the chemical and petroleum industries, uses those funds for the cleanup of abandoned and nonoperating hazardous waste sites, and authorizes the federal government to respond directly to the release or threatened release of substances that may pose a threat to human health or the environment.

_____ 8. Ash

    h. Solid waste collected by municipalities from households, small businesses, and institutions such as schools, prisons, municipal buildings, and hospitals.

_____ 9. Waste-to-energy

    i. The flow of solid waste that is recycled, incinerated, placed in a solid waste landfill, or disposed of in another way.

_____ 10. Hazardous waste

    j. Contaminated industrial or commercial sites that may require environmental cleanup before they can be redeveloped or expanded.

_____ 11. Superfund Act

    k. The residual nonorganic material that does not combust during incineration.

_____ 12. Brownfields

    l. An engineered ground facility designed to hold municipal solid waste (MSW) with as little contamination of the surrounding environment as possible.

MODULE 51: Waste Reduction Methods

Before You Read the Module

Focus on Learning Goals

Use the module learning goals to guide your reading. On a separate piece of paper, write down each goal and take notes to help you meet each learning goal. After studying this module, you should be able to:

- 51-1 describe the three Rs that divert materials from the waste stream.
- 51-2 explain how composting further reduces materials entering the waste stream.
- 51-3 explain how life-cycle analysis and integrated waste management reduce municipal solid waste.

Key Terms

Reduce, Reuse, Recycle (the three Rs)	Recycling	Life-cycle analysis (Cradle-to-grave analysis)
Source reduction	Closed-loop recycling	Integrated waste management
Reuse	Open-loop recycling	
	Composting	

While You Read the Module

Answer the following questions as you read. Use a separate sheet of paper if necessary.

Module 51: Waste Reduction Methods

The three Rs: reduce, reuse, recycle

1. Define "Reduce, Reuse, Recycle," also known as the three Rs.

Reduce

2. Describe why "reduce" is the first and best option.

3. What is "reduce" also known as?

4. Define source reduction.

5. Describe how source reduction can help with more than materials reduction.

6. Give an example of how a teacher could implement source reduction.

7. Describe how CDs used to be packaged. Describe two different methods of source reduction for CDs.

8. Give an example of how source reduction can be achieved by material substitution.

Reuse

9. Define reuse.

10. Describe some of the costs of reusing materials.

11. List other ways individuals can reuse materials.

12. Explain the pros and cons of reusing materials.

Recycle

13. Define recycling.

14. Define closed-loop recycling. Give an example.

15. Define open-loop recycling. Give an example.

16. Figure 51.4: Identify the years with the greatest growth of recycling.

17. Figure 51.5: Explain zero sort recycling programs. Identify other names of zero sort programs.

18. Describe the market difference for the variety of recycled materials. Give an example of how facilities compensate for the market differences in products.

19. Discuss the recycling controversy in New York City 20 years ago.

20. Explain the possible future consequences of rising recycling costs.

Composting reduces food and yard trimmings in the waste stream and enhances soil quality

21. Explain two problems with organic material ending up in landfills.

22. Define composting.

23. Explain why dairy and meat products are not typically added to compost.

24. Describe the different arrangements people can use for compost at home. Identify the basic "recipe."

25. Figure 51.7: Describe large scale composting facilities and how they function.

26. Identify two reasons that composting with worms might be a good option for home composting.

27. Identify why compost is a good soil nutrient.

New ways to think about municipal solid waste

Life-Cycle Analysis

28. Define life-cycle (cradle-to-grave) analysis.

29. Explain the limitations of life-cycle analysis. Give an example.

30. How can lifecycle analysis help a municipality make economic decisions about waste recycling?

Integrated Waste Management

31. Define integrated waste management.

32. Figure 51.9: Describe the significance of the horizontal and curved arrows.

33. Figure 51.9: List and describe improvements that lead to source reduction and waste reduction in the areas of: manufacturing, use, and waste.

34. Describe the philosophy of *Cradle to Cradle* from McDonough and Braungart.

35. Identify and explain two industries using new approaches to their products.

36. Describe upcycle.

Practice the Math: Calculating Recycling Rates

Read "Do the Math: Calculating Recycling Rates" on page 602. Try "Your Turn." For more math practice, do the following problems. Remember to show your work. Use a separate sheet of paper if necessary.

(a) According to Figure 51.4 on page 601 of the textbook, in 1990, approximately 25% of total MSW was recycled, resulting in a total 50 million metric tons recycled. Calculate the total MSW generated in 1990.

(b) According to Figure 51.4 on page 601 of the textbook, in 2010, approximately 33% of total MSW was recycled, resulting in a total 65 million metric tons recycled. Calculate the total MSW generated in 2010.

(c) Based on these two calculations, is the total MSW increasing or decreasing over time?

Review Key Terms

Match the key terms on the left with the definitions on the right.

_____ 1. Reduce, Reuse, Recycle (the three Rs)

    a. The process by which materials destined to become municipal solid waste (MSW) are collected and converted into raw materials that are then used to produce new objects.

_____ 2. Source reduction

    b. Recycling a product into the same product.

_____ 3. Reusing

    c. Using a product or material that would otherwise be discarded.

_____ 4. Recycling

    d. The breakdown of organic materials into organic matter (humus).

_____ 5. Closed-loop recycling

    e. An approach to waste disposal that employs several waste reduction, management, and disposal strategies to reduce their costs and reduce the environmental impact of MSW.

_____ 6. Open-loop recycling

    f. A popular phrase promoting the idea of diverting materials from the waste stream.

_____ 7. Composting

    g. A systems tool that examines the materials used and released throughout the lifetime of a product — from the product design and procurement of raw materials through their manufacture, use, and disposal

_____ 8. Life-cycle analysis (Cradle-to-grave analysis)

    h. Recycling one product into a different product.

_____ 9. Integrated waste management

    i. An approach to waste management that seeks to cut waste by reducing the use of potential waste materials in the early stages of design and manufacture.

MODULE 52: Sewage Treatment

Before You Read the Module

Focus on Learning Goals
Use the module learning goals to guide your reading. On a separate piece of paper, write down each goal and take notes to help you meet each learning goal. After studying this module, you should be able to:
- 52-1 describe the three major problems caused by wastewater pollution.
- 52-2 explain the modern technologies used to treat wastewater.

Key Terms

Biochemical oxygen demand (BOD)	Septic system	Leach field
	Septic tank	Manure lagoon
Cultural eutrophication	Sludge	
Fecal coliform bacteria	Septage	

While You Read the Module
Answer the following questions as you read. Use a separate sheet of paper if necessary.

Module 52: Sewage Treatment

There are three major problems caused by wastewater pollution

1. Identify and describe three reasons environmental scientists are concerned about wastewater pollutants.

Oxygen Demand

2. Describe the processes microbes require for decomposition.

3. Define biochemical oxygen demand (BOD).

4. Explain the difference between low BOD and high BOD. Give an example of each.

5. Describe dead zones and how they can be self-perpetuating.

Nutrient Release

6. Identify two limiting elements in an aquatic ecosystem.

7. Define cultural eutrophication.

8. Identify and list the chain of events in cultural eutrophication.

9. Describe the dead zone where the Mississippi River dumps into the Gulf of Mexico.

10. Identify how the number of dead zones around the world have changed from 1910 to 2018.

Disease-Causing Organisms

11. Figure 52.3: Describe two uses of the river seen in the photo.

12. What can wastewater carry that has possible detrimental effects?

13. Figure 52.4: Identify where cholera is common and why children are more susceptible.

14. Explain how hepatitis A has appeared in the United States more recently.

15. Identify the number of individuals in the world that do not have access to clean drinking water.

16. Identify the number of individuals worldwide who do not have access to proper sanitation and where these individuals are located.

17. Identify the number of children that die each year from diarrhea and why.

Monitoring for Wastewater Contamination

18. Define fecal coliform bacteria.

19. Describe *E. coli* and explain why it is considered an indicator.

20. When is human consumption of water not allowed?

21. When is fishing and swimming considered safe?

We have modern technologies used to treat wastewater

Septic Systems

22. Define septic system.

23. Define septic tank.

24. Define sludge.

25. Define septage.

26. Identify and describe the three layers that develop in a septic tank.

27. Define leach field.

28. List the steps that occur for septage movement and breakdown.

29. Identify an advantage and a disadvantage of septic systems.

Sewage Treatment Plants

30. Describe the most feasible locations for sewage treatment plants.

31. Explain primary treatment in sewage treatment plants.

32. Describe secondary treatment in sewage treatment plants.

33. Figure 52.7: Describe where sludge goes after primary and secondary treatment.

34. Explain why there is a need for tertiary treatment at sewage treatment plants.

35. Identify how wastewater in a sewage plant is disinfected after secondary treatment.

Legal Sewage Dumping

36. Explain why some developed countries pump raw sewage into rivers and lakes.

37. How big of a problem is legal sewage dumping?

38. Where is legal dumping causing the most concern?

39. Describe the impacts of raw sewage in the water.

40. Explain the solution to raw sewage in the water.

Animal Feed Lots and Manure Lagoons

41. Explain the reason for concern about manure from concentrated animal feeding operations.

42. Define manure lagoon.

43. Identify the benefit of manure lagoons.

44. Identify detrimental effects of manure lagoons.

Practice the Math: Building a Manure Lagoon

Read "Do the Math: Building a Manure Lagoon" on page 615. Try "Your Turn." For more math practice, do the following problem. Remember to show your work. Use a separate sheet of paper if necessary.

If an individual animal produces 50 L of manure each day and a manure lagoon needs to hold 45 days' worth of manure production for 1,500 cattle, what is the minimum capacity of the lagoon a farmer would need?

After You Read the Module

Review Key Terms
Match the key terms on the left with the definitions on the right.

_____ 1. Biochemical oxygen demand (BOD)

_____ 2. Cultural eutrophication

_____ 3. Fecal coliform bacteria

_____ 4. Septic system

_____ 5. Septic tank

_____ 6. Sludge

_____ 7. Septage

_____ 8. Leach field

_____ 9. Manure lagoon

a. A large, human-made pond lined with rubber built to prevent the manure from leaking into the groundwater.

b. A relatively small and simple sewage treatment system, made up of a septic tank and a leach field, often used for homes in rural areas.

c. A layer of fairly clear water found in the middle of a septic tank.

d. A component of a septic system, made up of underground pipes laid out below the surface of the ground

e. An increase in fertility in a body of water, the result of anthropogenic inputs of nutrients.

f. Solid waste material from wastewater.

g. The amount of oxygen a quantity of water uses over a period of time at a specific temperature.

h. A group of microorganisms that live in the intestines of humans, other mammals, and birds that serve as an indicator species for potentially harmful microorganisms associated with contaminated sewage.

i. A large container that receives wastewater from a house as part of a septic system.

MODULE 53: Lethal Dose 50% (LD$_{50}$) and Dose Response Curves

Before You Read the Module

Focus on Learning Goals

Use the module learning goals to guide your reading. On a separate piece of paper, write down each goal and take notes to help you meet each learning goal. After studying this module, you should be able to:

- 53-1 explain how dose-response curves are used to estimate lethal doses of chemicals.
- 53-2 identify how we estimate potential harm of chemicals in the environment.
- 53-3 describe the major philosophies of regulating chemicals in the environment.

Key Terms

Dose-response study	ED$_{50}$	Innocent-until-proven-guilty
Acute study	No-observed-effect level	principle
Chronic study	(NOEL)	Precautionary principle
LD$_{50}$	Environmental hazard	Stockholm Convention
Sublethal effect		REACH

While You Read the Module

Answer the following questions as you read. Use a separate sheet of paper if necessary.

Module 53: Lethal Dose 50% (LD$_{50}$) and Dose Response Curves

Scientists can determine the concentrations of chemicals that harm organisms

Dose-Response Studies

1. Define dose-response studies.

2. Explain how tadpoles might be used in a dose-response study.

3. Define acute study.

4. Define a chronic study.

5. Figure 53.2a: Explain the plotted data on the graph and the curve of the line.

6. Describe threshold dose.

7. Define LD_{50}.

8. Figure 53.2b: Explain LD_{50} on the graph.

9. Identify the goal of a chronic study.

10. Define sublethal effects.

11. Define ED_{50}.

12. Define no-observed-effect level (NOEL).

13. Figure 53.2b: Identify the NOEL value on the graph.

LD_{50} Studies

14. Describe the Toxic Substances Control Act of 1976.

15. Describe the Federal Insecticide, Fungicide, and Rodenticide Act of 1996.

16. Explain the method devised for testing chemicals in a way that represents the millions of species on Earth.

17. List the species tested for each category.

18. Identify the reason scientists now try to include testing on amphibians and reptiles.

19. Explain the calculation used for determining safe concentrations of substances for most animals.

20. Identify and explain the calculation used for obtaining safe concentrations of substances for humans.

We can estimate potential harm using risk assessment, risk acceptance, and risk management

Risk Assessment

21. Identify the two types of risk assessment.

22. Define an environmental hazard. Identify an example.

23. Describe a qualitative judgment and give an example.

24. Explain a quantitative risk assessment.

25. Figure 53.4: Identify and describe the difference in risk between an airplane accident and a motor vehicle accident. Which risk is higher?

26. Identify the equation for risk.

27. Explain the similarity in risk for dying from a plane crash and dying from consuming peanut butter.

28. Identify the factors the EPA used in the risk assessment for PCBs found in rivers from the 1940s to the 1970s.

29. Identify the EPA's two final results of the risk assessment on PCBs.

Risk Acceptance

30. Why is risk acceptance the most difficult of the three steps?

Risk Management

31. Identify what risk management integrates.

32. Describe the difference between people who implement risk assessments and those who conduct risk management.

33. Describe the final ruling of the maximum concentrations of arsenic in 2001. What factor determined the ruling?

Worldwide standards of risk are guided by two different philosophies

34. Define the innocent-until-proven-guilty principle.

35. Identify the downside of the innocent-until-proven-guilty principle.

36. Define the precautionary principle.

37. Identify who uses the precautionary principle.

38. Identify a country that uses the innocent-until-proven-guilty principle.

39. When were the first deaths from asbestos reported?

40. When were the first experiments conducted showing harmful effects from asbestos?

41. When did the European Union ban asbestos?

42. Describe what a study in the Netherlands showed would have happened if asbestos had been banned in 1965.

International Agreements on Hazardous Chemicals

43. Define Stockholm Convention.

44. Describe the "dirty dozen."

45. How many chemicals were on the list in 2017?

46. Define REACH.

47. What does REACH say about the responsibility of chemical manufacturers for their products?

Practice the Math: Estimating LD_{50} Values and Safe Exposures

Read "Do the Math: Estimating LD_{50} Values and Safe Exposures" on page 621. Try "Your Turn." For more math practice, do the following problem. Remember to show your work. Use a separate sheet of paper if necessary.

The LD_{50} of a particular pesticide for a rat is 2 mg/kg of mass. Assume that to determine the safe level of a pesticide for a dog you divide this LD_{50} value by 100. What amount of pesticide would be considered safe for a dog to ingest?

Review Key Terms

Match the key terms on the left with the definitions on the right.

_____ 1. Dose-response study

a. A 2001 agreement among 127 nations concerning 12 chemicals to be banned, phased out, or reduced.

_____ 2. Acute study

b. A principle based on the belief that when a hazard is plausible but not yet certain, we should take actions to reduce or remove the hazard.

_____ 3. Chronic study

c. A principle based on the belief that a potential hazard should not be considered an actual hazard until the scientific data definitively demonstrate that it actually causes harm.

_____ 4. LD_{50}

d. A study that exposes animals or plants to different amounts of a chemical and then looks for a variety of possible responses, including mortality or changes in behavior or reproduction.

_____ 5. Sublethal effect

e. The effective dose of a chemical that causes 50 percent of the individuals in a dose-response study to display a harmful, but nonlethal, effect.

_____ 6. ED_{50}

f. The lethal dose of a chemical that kills 50 percent of the individuals in a dose-response study.

_____ 7. No-observed-effect level (NOEL)

g. The highest concentration of a chemical that causes no lethal or sublethal effects.

_____ 8. Environmental hazard

h. An experiment that exposes organisms to an environmental hazard for a short duration.

_____ 9. Innocent-until-proven-guilty principle

i. A 2007 agreement among the nations of the European Union about regulation of chemicals; the acronym stands for registration, evaluation, authorization, and restriction of chemicals.

_____ 10. Precautionary principle

j. The effect of an environmental hazard that does not kill an organism but which may impair an organism's behavior, physiology, or reproduction.

_____ 11. Stockholm Convention

k. An experiment that exposes organisms to an environmental hazard for a long duration.

_____ 12. REACH

l. Anything in the environment that can potentially cause harm.

MODULE 54: Pollution, Human Health, Pathogens, and Infectious Diseases

Before You Read the Module

Focus on Learning Goals

Use the module learning goals to guide your reading. On a separate piece of paper, write down each goal and take notes to help you meet each learning goal. After studying this module, you should be able to:

- 54-1 describe how we establish cause and effect between pollutants and human health.
- 54-2 describe the different types of human diseases.
- 54-3 identify historic human pathogens that have cycled through the environment.
- 54-4 identify the major emergent infectious diseases in humans.
- 54-5 identify the laws that protect human health from pollutants and pathogens.

Key Terms

Retrospective study
Prospective study
Synergistic interaction
Disease
Infectious disease
Acute disease
Chronic disease
Epidemic
Pandemic
Dysentery
Plague
Malaria

Tuberculosis
Emergent infectious disease
Acquired Immune Deficiency
Syndrome (AIDS)
Human Immunodeficiency
Virus (HIV)
Ebola hemorrhagic fever
Mad cow disease
Prion
Swine flu
Bird flu

Severe acute respiratory
syndrome (SARS)
MERS-CoV
SARS-CoV-2
West Nile virus
Lyme disease
Zika virus disease
Clean Water Act
Safe Drinking Water Act
Maximum contaminant level
(MCL)

While You Read the Module

Answer the following questions as you read. Use a separate sheet of paper if necessary.

Module 54: Pollution, Human Health, Pathogens, and Infectious Diseases

We can establish cause and effect between pollutants and human health using retrospective and prospective studies

1. Describe the study of epidemiology.

Retrospective Studies

2. Define retrospective study.

3. Describe the initial effect of the 1984 Bhopal, India chemical accident on people in the area.

4. Describe the findings from the Bhopal, India retrospective studies.

Prospective Studies

5. Define prospective study.

6. Explain how a prospective study could be performed.

7. Define a synergistic interaction and give an example.

8. Describe the study of the effects of lead on children's intelligence.

Human diseases can be infectious or noninfectious

9. Figure 54.2(a): List the leading causes of death in the world from highest to lowest.

10. Define disease.

11. Define infectious disease and give an example.

12. Describe noninfectious diseases and give an example.

13. Identify the four types of infectious diseases that account for 80 percent of deaths from infectious diseases.

14. Define acute disease and give an example.

15. Define chronic diseases and give an example.

Risk Factors for Chronic Disease in Humans

16. Identify the top risk factors for chronic disease in low-income countries.

17. Identify the top risk factors for chronic disease in high-income countries.

18. Figure 54.4: Describe the trend through time on the graph for both modern risks and traditional risks.

Many pathogens have been historically important

19. Define epidemic.

20. Define pandemic.

21. Define dysentery and identify the leading cause.

Plague

22. Define plague.

23. What are the other names for the plague?

24. Identify the symptoms of the plague.

25. Figure 54.6: Describe the Black Death in Europe from the 1300s to the 1800s.

26. Describe a recent plague outbreak.

27. Describe the effective treatment of the plague.

Malaria

28. Define malaria.

29. Describe the stages and symptoms of malaria.

30. Identify the regions hardest hit with malaria.

31. Explain the origin and number of new malaria cases diagnosed in the United States every year.

32. Identify traditional approaches to combating malaria.

Tuberculosis

33. Define tuberculosis.

34. Describe how tuberculosis is spread from person to person.

35. Identify the symptoms of tuberculosis.

36. Identify the treatment of tuberculosis.

37. Describe why tuberculosis treatment is difficult and sometimes unsuccessful in developing countries.

38. Identify the regions that are becoming a concern for drug resistant tuberculosis.

Emergent infectious diseases pose new risks to humans

39. Define emergent infectious disease.

40. How often has an emergent infectious disease been observed?

41. Explain why emergent diseases are able to spread more rapidly now than in previous centuries.

HIV/AIDS

42. Define Acquired Immune Deficiency Syndrome (AIDS).

43. Define Human Immunodeficiency Virus (HIV).

44. Explain how the AIDS virus is spread.

45. Explain the possible origin of AIDS.

46. Explain the treatment for HIV.

Ebola Hemorrhagic Diseases

47. Define Ebola hemorrhagic fever.

48. Describe the symptoms of the Ebola virus.

49. Explain the treatment for the Ebola virus.

50. Describe the possible origin of the Ebola virus.

51. When was the first Ebola vaccine approved?

Mad Cow Disease

52. Define mad cow disease.

53. Define prion.

54. Figure 54.11: Describe the symptoms of mad cow disease in cattle.

55. Identify another name for mad cow disease.

56. How can mad cow disease be transmitted to humans? What is the disease known as?

57. Describe the processes by which mutant prions are transferred to other animals and humans.

58. Describe how people could be infected with mad cow disease but not show symptoms.

59. What new rules exist today in response to mad cow disease?

Swine Flu and Bird Flu

60. Define swine flu.

61. Describe the treatment of swine flu.

62. Define bird flu.

63. Explain the different reactions of wild and domesticated birds to infections.

64. Explain the detrimental effects of the H5N1 infections as of 2021.

SARS, MERS-CoV, and SARS-CoV-2

65. Define severe acute respiratory syndrome (SARS).

66. What are the symptoms of SARS similar to?

67. Define MERS-CoV.

68. Define SARS-CoV-2.

69. List the timeline of appearance of the SARS, MERS-CoV, and SARS-CoV-2 viruses and the estimated infected people and numbers of deaths.

70. Figure 54.13: Explain how the SARS-CoV-2 impacted the world differently from the other two coronaviruses.

West Nile Virus

71. Define West Nile virus.

72. Describe the symptoms of West Nile virus.

73. Figure 54.14: Explain how to reduce the occurrence of West Nile virus.

Lyme disease

74. Define Lyme disease.

75. Explain how a tick becomes the vector for Lyme disease.

76. Describe the symptoms of Lyme disease.

77. Identify the treatment for Lyme disease.

Zika virus disease

78. Define Zika virus disease.

79. Explain how Zika is transmitted.

80. Describe the symptoms of Zika virus disease and the associated risks.

81. Describe the treatment of the Zika virus.

Future Challenges to Human Health

82. Describe the way in which pathogens develop drug resistance.

83. Describe ways in which emergent diseases can be addressed in the future.

Laws exist to protect human health from pollutants and pathogens

The Clean Water Act

84. Define the Clean Water Act.

85. What does the Clean Water Act not include?

86. Identify more recent focuses of the Clean Water Act.

The Safe Drinking Water Act

87. Define the Safe Drinking Water Act.

88. Define maximum contaminant levels (MCL).

89. Table 54.1: Copy the maximum contaminant levels for drinking water shown in the table.

90. Table 54.2: Think about where you live; identify the waterways closest to you and what impairment may affect the water where you live.

91. What industry is exempt from the Safe Drinking Water Act?

Water Pollution Legislation in the Developing World

92. Explain why developing countries tend not to have water pollution legislation.

93. Describe the progress made in Brazil with the Tietê River.

Visual Representation 8: Water pollution in the Florida Everglades

94. Describe the negative effects of the building projects in the Florida Everglades and how they have altered the flow of water.

95. Identify how farmers are reducing the detrimental effects of agriculture on water.

96. Explain how oil exploration and drilling are impacting animal habitats.

97. Describe the initial mercury reduction in the Everglades.

Pursing Environmental Solutions

Purifying Water for Pennies

   98. How many people around the globe drink unsafe water?

   99. Identify the team that developed a solution to clean water effectively at a low cost. Describe the product.

   100. Describe the two-step process to cleaning water. How long does the process take?

   101. Explain how Procter & Gamble can make and distribute this product without profiting in the process.

   102. Describe how much water was purified in 2021 and the positive worldwide outcome. What is the goal by 2025?

   103. Describe how the purifying water packets can be useful in developed nations for people with access to clean water. How can this help the other countries with unsafe water?

Science Applied 8

Is Recycling Always Good for the Environment?

How do we begin to assess the benefits of recycling?

   104. Identify how to determine whether something should be recycled.

   105. Describe the life cycle of an aluminum can and the economic and environmental costs of each stage.

   106. Describe the life cycle of plastic containers and the economic and environmental costs of each stage.

107. Identify the percentage of aluminum cans manufactured in 2020 that were recycled, and the percentage of plastic bottles manufactured in 2018 that were recycled.

What other costs of recycling do we need to consider?

108. List other costs of recycling that we need to consider.

What other benefits of recycling do we need to consider?

109. Describe the benefits when a landfill can operate for longer.

After You Read the Module

Review Key Terms
Match the key terms on the left with the definitions on the right.

_____ 1. Retrospective study

_____ 2. Prospective study

_____ 3. Synergistic interaction

_____ 4. Disease

_____ 5. Infectious disease

_____ 6. Acute disease

_____ 7. Chronic disease

_____ 8. Epidemic

_____ 9. Pandemic

_____ 10. Dysentery

_____ 11. Plague

a. Any impaired function of the body with a characteristic set of symptoms.

b. A disease that rapidly impairs the functioning of a person's body.

c. A situation in which a pathogen causes a rapid increase in disease.

d. A coronavirus that causes the disease known as Middle Eastern Respiratory Syndrome.

e. Legislation that supports the "protection and propagation of fish, shellfish, and wildlife and recreation in and on the water" by maintaining and, when necessary, restoring the chemical, physical, and biological properties of surface waters.

f. An epidemic that occurs over a large geographic region, such as an entire continent.

g. A virus that lives in hundreds of species of birds and is transmitted among birds by mosquitoes.

h. A type of flu caused by the H1N1 virus.

i. An infectious disease with high death rates, caused by several species of Ebola viruses.

j. A highly contagious disease caused by the bacterium Mycobacterium tuberculosis that primarily infects the lungs.

k. A disease caused by a bacterium (*Borrelia burgdorferi*) that is transmitted by ticks.

_____ 12. Malaria

l. A disease caused by a pathogen.

_____ 13. Tuberculosis

m. An infectious disease caused by one of several species of protists in the genus *Plasmodium*.

_____ 14. Emergent infectious disease

n. A disease that slowly impairs the functioning of a person's body.

_____ 15. Acquired Immune Deficiency Syndrome (AIDS)

o. A coronavirus that causes the disease known as Covid-19

_____ 16. Human Immunodeficiency Virus (HIV)

p. A type of flu caused by a coronavirus.

_____ 17. Ebola hemorrhagic fever

q. A type of flu caused by the H5N1 virus.

_____ 18. Mad cow disease

r. A disease caused by a pathogen that causes fetuses to be born with unusually small heads and damaged brains.

_____ 19. Prion

s. A type of virus that causes Acquired Immune Deficiency Syndrome (AIDS).

_____ 20. Swine flu

t. A study that monitors people who might become exposed to an environmental hazard, such as a harmful chemical, at some time in the future.

_____ 21. Bird flu

u. A situation in which two risks together cause more harm than expected based on the separate effects of each risk alone.

_____ 22. Severe acute respiratory syndrome (SARS)

v. The standard for safe drinking water established by the EPA under the Safe Drinking Water Act.

_____ 23. MERS-CoV

w. A disease in which prions mutate into deadly pathogens and slowly damage a cow's nervous system.

_____ 24. SARS-CoV-2

x. An infectious disease caused by a bacterium (*Yersinia pestis*) that is carried by fleas.

_____ 25. West Nile virus

y. An infectious disease that has not been previously described or has not been common for at least the prior 20 years.

_____ 26. Lyme disease

z. A study that monitors people who have been exposed to an environmental hazard, such as a harmful chemical, at some time in the past.

_____ 27. Zika virus disease

aa. An infection of the intestines that causes diarrhea, which results in dehydration and can cause death.

_____ 28. Clean Water Act

bb. Legislation that sets the national standards for safe drinking water.

_____ 29. Safe Drinking Water Act

cc. A small, beneficial protein that occasionally mutates into a pathogen.

_____ 30. Maximum contaminant level (MCL)

dd. An infectious disease caused by the human immunodeficiency virus (HIV).

UNIT 8 Review Exercises

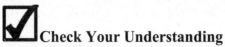

Check Your Understanding

Review "Learning Goals Revisited" on pages 561, 574, 582, 596, 607, 616, 626, and 644 of your textbook. Compare the notes you took while reading each module. Complete these exercises to review the chapter. Use a separate piece of paper if necessary.

1. What are the three reasons that environmental scientists are concerned about human wastewater as a pollutant?

2. Describe the possible result to an organism of high pesticide exposure.

3. Identify the environmental effect of an oil spill on the ocean floor.

4. List and describe ways that humans alter and control the movement of water.

5. Explain how bioaccumulation is different from biomagnification.

6. Describe the causes and consequences of eutrophication.

7. Describe the environmental consequences of sanitary landfills and incineration of waste.

8. Explain the RCRA, CERCLA, and the Brownfields Program.

9. List and describe the four reduction methods of solid waste.

10. Summarize the steps of a sewage treatment plant.

11. Explain the difference between LD_{50} and ED_{50}.

12. Complete the following chart:

Pathogens	**How Disease is Spread**
Plague	
Malaria	
Tuberculosis	
HIV/AIDS	
SARS	
MERS CoV	
SARS Co-V-2	
West Nile	
Zika	
Cholera	

Practice for Free-Response Questions

Complete this exercise to build and practice the skills you will need to answer free-response questions on the exam. Use a separate sheet of paper if necessary.

1. Oil and other petroleum products can enter the oceans as spills from oil tankers or pipelines.

   (a) Identify and describe a physical barrier used to remediate oil pollution.

   (b) Identify and describe a chemical method to remediate oil pollution.

   (c) Identify and describe a biological approach to remediate oil pollution.

2. In the United States, most electronic devices are not designed to be easily dismantled after they are discarded.

   (a) Identify one product that would be considered electronic waste, also known as e-waste.

   (b) Describe a negative environmental or health effect from the throwing away or recycling e-waste.

3. Humans have impacted waterbodies through a variety of activities and this is more apparent in wetlands, standing water or water-saturated soil during some part of the year.

   (a) Name one type of wetlands.

   (b) Identify an ecosystem service of a wetland.

   (c) Explain one reason for humans to alter wetlands.

Unit 8 Multiple-Choice Review Exam

1. Water diversion has split which body of water and reduced its surface area by 60 percent?
   (a) Lake Victoria
   (b) The Caspian Sea
   (c) Lake Baikal
   (d) The Aral Sea

2. Which part of the world produces 50 percent of the world's desalinated water?
   (a) North America
   (b) Middle East
   (c) Europe
   (d) South America

3. What is the purpose of a fish ladder?
   (a) They allow migrating fish to get around a dam.
   (b) They serve as a location for breeding.
   (c) They allow people to transport fish.
   (d) They prevent invasive species from entering a reservoir.

4. Desalinization is
   (a) used extensively across Europe.
   (b) a problem with saltwater intrusion.
   (c) occurring at a rapid rate due to over irrigating farmland.
   (d) helping water-poor countries obtain fresh water.

5. Which is an example of a point source water pollutant?
   (a) animal feedlots
   (b) runoff from parking lots
   (c) factory waste water
   (d) residential lawns

6. When sewage contaminates a body of water, it can lead to a lower dissolved oxygen level in the water because
   (a) the sewage has put a high BOD on the water.
   (b) the water has too many fish and other organisms.
   (c) the sewage magnifies existing toxins.
   (d) of eutrophication.

7. The World Health Organization estimates that _____ people do not have access to sufficient supplies of safe drinking water.
   (a) 1 out of every 50
   (b) 1 out of every 25
   (c) 1 out of every 10
   (d) 1 out of every 4

Match the following steps in the sewage treatment process.

8. _____ Biological   (a) Solid waste materials settle out

9. _____ Chemical   (b) Bacteria break down 85-90 percent of organic matter

10. _____ Mechanical (c) Chlorine, ozone, or ultraviolet light are used

11. Which makes up most of the municipal solid waste stream in the United States?
    (a) paper
    (b) plastic
    (c) yard trimmings
    (d) wood

12. Which uses the least amount of energy?
    (a) reduce
    (b) reuse
    (c) closed-loop recycling
    (d) incineration

13. Which is an advantage of incineration?
    (a) production of ash
    (b) releases carbon dioxide and water
    (c) reduction of waste volume or mass
    (d) release of heat

14. Integrated waste management
    (a) requires more time and money than other waste management.
    (b) adjusts manufacturing to allow for future recycling.
    (c) increases material used in packaging designs.
    (d) Encourages the use of single use items for faster disposal.

15. Which is an infectious disease?
    (a) asbestosis
    (b) heart disease
    (c) tuberculosis
    (d) mesothelioma

16. Which is a neurotoxin?
    (a) arsenic
    (b) polychlorinated biphenyls
    (c) lead
    (d) synthetic hormones

17. Which experiment results correctly describes the effects of a LD_{50} dose of a chemical?
    (a) One out of every 50 rats die.
    (b) Fifty out of every 100 rats die.
    (c) One out of every 50 rats gets sick.
    (d) Fifty out of every 100 rats get sick.

18. Which would be an example of a synergistic interaction?
    (a) Increased riders in a car leads to increased deaths.
    (b) Smoking causes lung disease.
    (c) Tuberculosis is spread among the poor.
    (d) A smoker who is also exposed to asbestos is more likely to get lung cancer.

19. The chemical DDT consumed by plankton is transferred to small fish, to larger fish, and eventually to predatory birds. This is an example of
    (a) a food chain.
    (b) a synergistic effect.
    (c) biomagnification.
    (d) teratogens.

20. A scientist is studying a local river and notices that many of the male fish, reptiles, and amphibians are becoming feminized; males possess testes that have low sperm counts, and in some cases, testes that produce both eggs and sperm. Based on this evidence, what might the scientist suspect?
    (a) The pH of the river has changed because of acid precipitation.
    (b) A local factory has been dumping wastewater with high levels of PCBs.
    (c) The river contains a large quantity of gasoline runoff from a local gas station.
    (d) The organisms have been exposed to wastewater containing pharmaceutical drugs that mimic estrogen.

21. Factors included in a risk assessment include:
    (a) industrial research, identify the hazard, find "safe" consumer product
    (b) identify the hazard, determine the acceptable level of risk, determine policy from citizens
    (c) identify the hazard, change in purchasing habits, determine policy from industry
    (d) changes in packaging design, recovery for recycling, identify the hazard

22. The Safe Drinking Water Act
    (a) allows the contaminant of fecal coliform to be at 10 ppb.
    (b) changed to allow arsenic to be measured at 50 ppb.
    (c) protects fish, shellfish, and wildlife.
    (d) includes both surface and groundwaters.

Match the following chemicals to the type of concern.

23. _____ Alcohol   (a) Endocrine disruptor

24. _____ Mercury   (b) Teratogen

25. _____ Phthalates   (c) Neurotoxin

26. Aquatic ecosystems harmed by eutrophication are due to
    (a) high levels of sedimentation.
    (b) low nutrient levels.
    (c) water runoff from agriculture fields.
    (d) low of levels of algae.

27. High water temperatures due to thermal pollution
    (a) can lead to high BOD and high dissolved oxygen levels.
    (b) has very little effect on water organisms.
    (c) can kill many species through desalination shock.
    (d) can lead to low levels of dissolved oxygen.

28. The United States EPA identifies four characteristics of hazardous waste, including
    (a) corrosivity, toxicity, pathogens, and teratogens.
    (b) reactivity, ignitability, leachate, and toxicity.
    (c) ignitability, reactivity, corrosivity, and toxicity.
    (d) septage, leachate, toxicity, and fecal coliform.

29. Composting, the breakdown of organic materials into humus, is best managed by
    (a) allowing the piles to sit, undisturbed for long periods of time.
    (b) adding outdated or expired meats and dairy.
    (c) keeping the pile as dry as possible.
    (d) layering dry materials, dried cut grass, with wet materials, kitchen vegetables.

30. One of the best-known hazardous waste landfills cleaned up by Superfund is
    (a) Cuyahoga River.
    (b) Hudson River.
    (c) Aral Sea
    (d) Love Canal.

UNIT 9

Global Change

Unit Summary

This unit looks at three processes: global change, global climate change, and global warming. The unit carries the most weight of any unit on the AP® Environmental Science Exam, and for good reason! The unit ties together many of the themes that have developed throughout the book, and focuses heavily on Science Practice 7, Environmental Solutions. of the topics covered are particularly engaging: ocean warming and acidification, impacts of climate change, and loss of biodiversity. While reading, be prepared to look back on previous units to make connections.

MODULES IN THIS UNIT

Module 55: Stratospheric Ozone Depletion and Its Reduction
Module 56: The Greenhouse Effect
Module 57: Increases in the Greenhouse Gases and Global Climate Change
Module 58: Ocean Warming and Ocean Acidification
Module 59: Invasive Species, Endangered Species, and Human Impacts on Biodiversity

---

Unit Opening Case: *Sea Turtle Responses to a Warming World*

This case study looks at how rising global temperatures have serious consequences for animal life. The sex of a green sea turtle is determined by temperature and rising sea temperatures have caused green sea turtles to be born disproportionately female. Researchers have discovered that in the past 25 years the subadult turtles that hatched in the warmer northern beaches around the Great Barrier Reef are 99.8 percent female, as opposed to their southern, cold-water counterparts which are 65-69 percent female. These biased sex ratios have serious consequences because the female turtles will not be able to find mates when they come of breeding age. Temperature-based sex determination is not exclusive to green sea turtles, and similar problems can be found with other species. For example, certain species of lizards are approaching 70 percent male and the reduction in the female population has made it difficult for males to find mates.

Do the Math

This unit contains the following "Do the Math" boxes to help prepare you for calculation questions you might encounter on the exam.

- "Projecting Future Increases in CO_2" (page 677)
- "Projecting Future Declines in Sea Ice" (page 683)
- "Estimating Percentages in the Conservation Status of North American Mammals" (page 698)

To make sure you understand the concepts and techniques presented in these boxes, do the practice problems presented in the text as well as the additional "Practice the Math" problems that appear in Module 57 and Module 59 of this study guide.

MODULE 55: Stratospheric Ozone Depletion and Its Reduction

Before You Read the Module

Focus on Learning Goals
Use the module learning goals to guide your reading. On a separate piece of paper, write down each goal and take notes to help you meet each learning goal. After studying this module, you should be able to:
- 55-1 explain how stratospheric ozone forms and the benefits it provides.
- 55-2 identify the cause of depleted stratospheric ozone.
- 55-3 describe efforts made to reduce ozone depletion.

Key Terms

Chlorofluorocarbons (CFCs)
Montreal Protocol

While You Read the Module
Answer the following questions as you read. Use a separate sheet of paper if necessary.

Unit Opening Case: Sea Turtle Responses to a Warming World

1. Describe the shocking discovery about green sea turtles.

2. Explain how the genetic markers were useful to the researchers.

3. What difference did researchers find between the southern and northern female green sea turtle populations?

4. Explain the significance of the 25-year marker.

5. How did these turtles produce so many females?

6. Identify why the change in sex ratio is a concern for the future of the green sea turtle over the next two decades.

7. How is the sex ratio of the tuatara lizard different from the green sea turtle?

8. Explain how the mating behavior of the tuatara could be an additional detriment to their population.

9. Identify a prediction for the tuatara lizard's future.

Module 55: Stratospheric Ozone Depletion and Its Reduction

10. Explain the difference between tropospheric ozone and stratospheric ozone.

11. Figure 55.1: Describe which two types of ultraviolet radiation are altered by the stratospheric ozone layer.

Stratospheric ozone is beneficial to the health and survival of life on Earth

Formation of Stratospheric Ozone

12. List and describe the steps of ozone formation in the stratosphere.

Benefits of Stratospheric Ozone

13. List the three types of ultraviolet wavelengths and their individual characteristics.

14. Develop your own phrase or rhyme to remember the difference between tropospheric ozone and stratospheric ozone.

Humans have depleted the ozone layer through the use of CFCs

The Chemical Reaction Between CFCs and Ozone

15. Define chlorofluorocarbons (CFCs) and identify their uses.

16. When were CFCs introduced?

17. Why were CFCs considered safe?

18. Explain the adverse effect of CFCs on the stratosphere.

19. List and describe the steps in the process of ozone breakdown from CFCs.

20. What is a role of a single chlorine molecule in the process of ozone breakdown from CFCs?

CFC Destruction of the Ozone Layer

21. When and where did researchers notice the depletion of the ozone layer?

22. Figure 55.2: Describe the overall trend of ozone concentrations from 1980 to 2020.

23. Describe the "ozone hole."

24. Identify the time of year ozone depletion occurs in the Antarctic.

25. Figure 55.3: Identify where the increases of UV radiation occurred.

26. What impact has the increase in UV-B radiation had on people from countries near the Antarctic ozone hole?

Nations around the world have agreed to reduce CFC production to reduce ozone depletion

27. Describe the Montreal Protocol.

28. How many countries signed to the Montreal Protocol in 1996?

29. Describe one of the alternatives to CFCs and its possible detrimental effects.

30. Explain recent trends of CFC concentration and ozone depletion.

After You Read the Module

Review Key Terms
Match the key terms on the left with the definitions on the right.

1. Chlorofluorocarbons (CFCs)   a. A commitment by 24 nations to reduce CFC production by 50 percent by the year 2000.

2. Montreal Protocol   b. Chemical that can be used for cooling refrigerators and air conditioners.

MODULE 56: The Greenhouse Effect

Before You Read the Module

Focus on Learning Goals
Use the module learning goals to guide your reading. On a separate piece of paper, write down each goal and take notes to help you meet each learning goal. After studying this module, you should be able to:
- 56-1 describe how we can distinguish between global change, global climate change, and global warming.
- 56-2 explain the process underlying the greenhouse effect.
- 56-3 identify the sources of greenhouse gases.

Key Terms

Global change
Global climate change

Greenhouse effect

Greenhouse warming
potential (GWP)

While You Read the Module

Answer the following questions as you read. Use a separate sheet of paper if necessary.

Module 56: The Greenhouse Effect

Global change includes global climate change and global warming

1. Define global change.

2. Figure 56.1: List two examples of a global change.

3. Describe the global change of mercury concentrations and the detrimental effects.

4. Define global climate change.

5. Figure 56.1: List the examples of global climate change.

6. Figure 56.1: List the examples of global warming.

Solar radiation and greenhouse gases make our planet warm

The Sun–Earth Heating System

7. Identify and describe the two types of heat that affect the Earth's warmth.

8. Figure 56.2: Identify the colors that represent incoming solar radiation and infrared radiation.

9. Figure 56.2: List the processes that occur in the atmosphere that cause the planet to warm.

10. Define greenhouse effect.

11. Explain the steady state of the Sun-Earth heating system.

12. Describe factors that influence short-term fluctuations of the Sun-Earth heating system.

Gases That Cause the Greenhouse Effect

13. List the greenhouse gases.

14. Explain the atmospheric location for ozone as a greenhouse gas.

15. Identify the greenhouse gas that is human made.

16. Describe what the temperature of Earth would be without any greenhouse gases.

17. Define greenhouse warming potential (GWP).

18. Table 56.1: Copy the table.

19. Which greenhouse gas has the greatest greenhouse warming potential and why?

Sources of greenhouse gases are both natural and anthropogenic

Natural Sources of Greenhouse Gases

20. Identify the gas associated with volcanic eruptions.

21. Describe the short-term climatic effects of a volcanic eruption.

22. Identify which gas is released during decomposition and digestion. Give examples of each.

23. Identify a natural greenhouse gas that is formed as part of the nitrogen cycle and describe how it is formed.

24. Explain the significance of water vapor as a greenhouse gas.

Anthropogenic Sources of Greenhouse Gases

25. Explain how fossil carbon formed.

26. List the fossil fuels that produce carbon dioxide from greatest amount to the least.

27. List the other greenhouse gases that are released during the production of fossil fuels.

28. Explain how particulate matter could play a role in global warming.

29. List and describe four agricultural practices and their associated greenhouse gases.

30. Explain how forest destruction can lead to higher levels of greenhouse gases.

31. Describe how landfills produce methane.

32. Describe the gases used as alternatives to CFCs and how they compare. Explain the future use of these gases.

33. Figure 56.6a: Identify the top two contributors to methane gas release in the United States and the percentages.

34. Figure 56.6b: Identify the top contributor to nitrous oxide gas release in the United States and the percentage.

35. Figure 56.6c: Identify the top two contributors to carbon dioxide gas release in the United States and the percentages.

After You Read the Module

Review Key Terms
Match the key terms on the left with the definitions on the right.

_____ 1. Global change

a. A type of global change that is focused on changes in the average weather that occurs in an area over a period of years or decades.

_____ 2. Global climate change

b. Absorption of infrared radiation by atmospheric gases and reradiation of the energy back toward Earth.

_____ 3. Greenhouse effect

c. Change that occurs in the chemical, biological, and physical properties of the planet.

_____ 4. Greenhouse warming potential (GWP)

d. An estimate of how much a molecule of any compound can contribute to global warming over a period of 100 years relative to one molecule of CO_2.

MODULE 57: Increases in the Greenhouse Gases and Global Climate Change

Before You Read the Module

Focus on Learning Goals
Use the module learning goals to guide your reading. On a separate piece of paper, write down each goal and take notes to help you meet each learning goal. After studying this module, you should be able to:
- 57-1 explain how CO_2 concentrations have changed over the past 7 decades.
- 57-2 describe how temperatures and greenhouse gases have varied historically.
- 57-3 describe how global climate change has affected the environment.
- 57-4 explain how global climate change is affecting populations.

While You Read the Module
Answer the following questions as you read. Use a separate sheet of paper if necessary.

Module 57: Increases in the Greenhouse Gases and Global Climate Change

CO_2 concentrations have increased for the past 7 decades

1. Describe the Intergovernmental Panel on Climate Change (IPCC).

2. Explain the mission of the Intergovernmental Panel on Climate Change (IPCC).

Measuring CO_2 Concentrations

3. Describe what scientists in the first half of the 20th century believed about what would happen to excess carbon dioxide released into the atmosphere.

4. How did Charles David Keeling change the way atmospheric carbon dioxide was tested?

5. Describe the causes of seasonal variations of carbon dioxide seen in the red line in Figure 57.1.

6. Figure 57.1: Describe the trend of atmospheric carbon dioxide from 1954 to 2020.

CO₂ Emissions Among Nations

7. Explain the current production of carbon dioxide emissions by developed and developing countries.

8. Describe the amount of world of carbon dioxide emissions that come from China.

9. Describe the amount of world of carbon dioxide emissions that come from the United States.

10. Figure 57.2: Explain the difference between the positions of China and India in (a) and (b).

Global temperatures and greenhouse gases have dramatically increased over the past two centuries

11. Describe the global temperature data from 1880 through 2020.

12. Explain how 1.1°C (2.0°F) is considered a substantial increase for the average global temperature.

13. How can scientists determine if recent global changes in temperature are typical?

14. Describe how scientists use foraminifera to gather information about temperature changes.

15. Describe the significance of air bubbles being trapped in ice.

16. Figure 57.5: Describe the pattern of atmospheric CO_2 from the last 400,000 years.

17. Figure 57.6: Identify two other greenhouse gases that have shown an increase in concentration through time.

18. Figure 57.6: Identify the trend in concentrations of greenhouse gases.

19. Identify the possible cause of the increase in greenhouse gases since the 1880s.

Climate Models Predicting Future Global Temperatures

20. Identify the tool that can predict future climate change.

21. Explain how computer models can accurately reflect today's climate.

22. Describe the confidence levels in computer models for temperatures compared to climate models.

23. Figure 57.7: Describe the highest and lowest temperature changes that could occur worldwide based on model estimates.

24. List possible predictions for the future as temperatures increase.

Global climate change is causing ice to melt and sea levels to rise

Melting Polar Ice Caps

25. Figure 57.8: Describe the amount of decline of Arctic ice from 1979 to 2021.

26. Describe the amount of ice lost in Antarctica and Greenland from 2000 to 2021.

Melting Glaciers

27. Describe the effect melting has had on Glacier National Park in northwest Montana.

28. Describe the detrimental effects that could occur if the glaciers are lost.

Ocean Currents

29. Describe the movement of the thermohaline circulation.

30. Explain how the thermohaline circulation could shift with more melting ice.

31. Identify the effect of a thermohaline circulation shutdown on Europe.

Warming Soils and Permafrost

32. List and describe the detrimental effects of melting permafrost.

Rising Sea Levels

33. Describe two ways the rise in global temperatures is affecting sea levels.

34. Identify scientific predictions for the amount that the sea level could rise.

35. Identify the detrimental effects of a rise in sea levels to coastal cities and low-lying islands.

36. Explain the possible effects on aquatic habitats of sea levels rising.

Global climate change is affecting the timing and performance of plants and animals

37. Explain the conclusions the IPCC arrived at regarding plant growing seasons after having reviewed approximately 2,500 scientific papers.

38. List the effects scientists have seen due to the extended growing seasons.

39. List the potential effects of climate change on humans, their health, and agriculture.

40. Identify how animals have historically responded to climate changes.

41. Explain the difficulty in migration today.

42. Describe how temperature change has affected the food source of the pied flycatcher.

43. Identify how much the pied flycatcher population has declined.

44. Describe the food web of Arctic polar bears.

45. Describe how seal hunting has changed for polar bear populations. Explain the detrimental effect this has had on polar bears.

46. What new opportunities might be available because of melting sea ice?

Practice the Math: Projecting Future Increases in CO_2

Read "Do the Math: Calculating the Magnitude of Oil Pollution" on page 561. Try "Your Turn." For more math practice, do the following problem. Remember to show your work. Use a separate sheet of paper if necessary.

From 1960 to 2010, the concentration of CO_2 in the atmosphere increased from 320 to 390 ppm. Using 2010 as your starting point, if the annual rate of CO_2 increase is 1.4 ppm, what concentration of CO_2 do you predict for the year 2150?

Practice the Math: Projecting Future Declines in Sea Ice

Read "Do the Math: Calculating the Magnitude of Oil Pollution" on page 561. Try "Your Turn." For more math practice, do the following problems. Remember to show your work. Use a separate sheet of paper if necessary.

As we calculated in "Do the Math," there were 5.27 million km^2 of polar ice remaining in 2009.

(a) If the polar melt increases from thirteen percent to 14 percent per decade, how much polar sea ice will there be in 2019, 2029, and 2039?

(b) If the polar melt increases to 14 percent per decade, what percentage of the original 1979 ice (8 million km^2) would remain in 2039? (Round to two significant figures.)

MODULE 58: Ocean Warming and Ocean Acidification

Before You Read the Module

Focus on Learning Goals

Use the module learning goals to guide your reading. On a separate piece of paper, write down each goal and take notes to help you meet each learning goal. After studying this module, you should be able to:

- 58-1 describe how ocean warming is altering ocean ecosystems.
- 58-2 explain how climate change affects ocean pH.
- 58-3 explain the international agreements on global climate change.

Key Terms

Ocean acidification   Paris Climate Agreement
Kyoto Protocol   (Paris Climate Accord)

While You Read the Module

Answer the following questions as you read. Use a separate sheet of paper if necessary.

Module 58: Ocean Warming and Ocean Acidification

Ocean warming is affecting a variety of marine species

1. Describe how lake and ocean temperatures have changed in the last four decades.

2. How much heat have the oceans absorbed from the excess created by humans?

Impacts on Marine Species

3. Identify and explain how researchers determined that marine species were shifting their distribution in the North Sea due to temperature changes.

4. What is the economic impact of fish species moving to the North Sea?

5. Identify the impact that an increase of 1 C° in water temperature can have on corals.

6. Describe coral bleaching and the underlying causes.

7. Explain what happens if coral bleaching lasts more than a few weeks.

Increasing CO_2 concentrations are causing ocean acidification

8. Figure 58.2: Reproduce this diagram in your notes. Explain the chemical reaction that is occurring.

9. Define ocean acidification.

10. Explain why ocean acidification can be harmful to a wide variety of species.

11. How can ocean acidification affect our food supply?

International agreements include the Kyoto Protocol and the Paris Agreement

12. Define the Kyoto Protocol.

13. List the different levels of emission restrictions by country.

14. Describe the purpose of developed countries paying the most for emissions control.

15. Describe the two options to stabilize greenhouse gases.

16. Identify a method for carbon sequestration.

17. Explain the stance the United States took with the Kyoto Protocol and the reason for its decision.

18. Describe the developments in fuel efficiency regulations in the United States between 2011 and 2020.

19. List the countries, according to the World Bank, that have changed their emission of greenhouse gases from 1991 to 2018.

20. Define the Paris Climate Agreement (Paris Climate Accord).

21. Explain how each country will help reduce greenhouse gases according to the Paris Climate Agreement.

\

After You Read the Module

Review Key Terms
Match the key terms on the left with the definitions on the right.

_____ 1. Ocean acidification

      a. An international agreement that sets a goal for global emissions of greenhouse gases from all industrialized countries to be reduced by 5.2 percent below their 1990 levels by 2012.

_____ 2. Kyoto Protocol

      b. A pledge by 196 countries to keep global warming less than 2 C ° above pre-industrial levels.

_____ 3. Paris Climate Agreement (Paris Climate Accord)

      c. A process in which an increase in ocean CO_2 causes more CO_2 to be converted to carbonic acid, which lowers the pH of the water.

MODULE 59: Invasive Species, Endangered Species, and Human Impacts on Biodiversity

Before You Read the Module

Focus on Learning Goals

Use the module learning goals to guide your reading. On a separate piece of paper, write down each goal and take notes to help you meet each learning goal. After studying this module, you should be able to:

- 59-1 identify the threat posed by invasive species.
- 59-2 describe why species are becoming endangered.
- 59-3 identify how human activities are affecting genetic biodiversity.
- 59-4 explain the causes of declining biodiversity.
- 59-5 identify how we can conserve biodiversity.

Key Terms

Endangered species
Lacey Act
Marine Mammal Protection Act
Endangered Species Act

Convention on International Trade in
Endangered Species of Wild Fauna and Flora
(CITES)

While You Read the Module

Answer the following questions as you read. Use a separate sheet of paper if necessary.

Module 59: Invasive Species, Endangered Species, and Human Impacts on Biodiversity

Invasive species can negatively impact native species, ecosystems, and human activities

1. List the characteristics of native, exotic, and invasive species.

2. Discuss the benefits and detrimental effects of the zebra mussels in the Great Lakes ecosystems.

3. How did honeybees arrive in the United States? Describe the negative effects.

4. Why was the kudzu vine introduced in the United States in 1876?

5. Explain how the kudzu vine became an invasive species.

6. Describe the two concerns about the effects of silver carp in the ecosystem.

Controlling Invasive Species

7. Explain the process of the introduction of the prickly pear cactus from South America to Australia, and how it was controlled.

8. What is the most effective way of dealing with invasive species?

9. Give an example of the current practices to reduce the introduction of exotic species.

Plants and animals are becoming endangered due to human activities

10. Explain why the current mass extinction is unique.

The Percentage of Endangered Plants and Animals

11. Describe extinct species.

12. Define endangered species.

13. Describe threatened species.

14. Describe near-threatened species.

15. Describe least-concern species.

16. Describe data-deficient species.

17. What organization helps the world understand the current loss of species?

18. Figure 59.3: List plant and animal groups in order from lowest to highest percentages of threatened or endangered species.

19. Figure 59.3: Identify which two groups have the largest percentage of species of least concern.

20. Identify a challenge in evaluating the status of different organisms.

Causes of Endangered Species

21. List the reasons for the decline in the abundance of species.

Human activities are causing declines in genetic biodiversity

22. Figure 59.4: Explain the decline and increase in genetic diversity of the Florida panther.

23. Explain the human activity that led to the decline of the Florida panther.

24. Describe the number of domesticated animal species that humans have bred and the reason for doing so.

25. Figure 59.6: Identify which domestic breeds have experienced the greatest number of extinctions.

26. Figure 59.6: Identify the domesticated breeds with the greatest risk for extinction.

27. Identify a possible reason that crop plants have lost their genetic diversity.

28. Identify the number of crops that have been historically cultivated and how many varieties have been lost over the last century.

29. Describe what happened when farmers in Ireland in the 1840s planted one variety of potato.

30. Explain how the nations of the world have responded to the declining seed diversity.

31. Describe the international seed vault located in Norway. Explain the significance of the location.

32. What is the capacity of the Svalbard Global Seed Vault and how many seed samples are stored there?

Declining biodiversity has a wide variety of causes

33. Explain the acronym used to remember the threats of declining biodiversity.

Habitat Destruction

34. Identify the primary causes of habitat loss.

35. Describe the habitat of the northern spotted owl.

36. Describe why the northern spotted owl experienced habitat loss and how this has affected its population.

37. Describe how the United States has compensated for the deforestation that occurred during the 1700s and 1800s.

38. Identify other habitats that are in danger of being lost and the percent of loss according to the Millennium Ecosystem Assessment.

39. Describe the cause for the decline of coral.

40. Describe how a reduced habitat has a detrimental effect on North American songbirds.

Invasive Species and Human Population Growth

41. How can invasive species affect native species?

42. Identify how human population growth impacts nature.

Pollution

43. List the pollutants or pollutant types that threaten biodiversity.

44. Identify the two pollutants responsible for the loss of many animals in the BP oil spill.

Climate Change

45. Explain the two responses different species may have to changes in climate.

46. Explain why the plants in a southwestern Australia peninsula may decline.

Overexploitation

47. Explain overexploitation and provide examples.

48. Describe two factors that caused the dodo bird to become extinct.

49. Identify the number of bison lost to overharvesting.

50. Describe how the passenger pigeon became extinct.

To conserve biodiversity, we need to reduce the threats

Conservation of Single Species

51. What can be done to help a single species recover from decline?

52. Explain the reasoning for a species to be captured and placed into a captive breeding program and the expected outcome.

53. Describe the successful captive breeding program of the California condor.

54. Define the Lacey Act.

55. Define the Marine Mammal Protection Act.

56. List some of the mammals that have been included by the Marine Mammal Protection Act.

57. Describe how the Endangered Species Act assists in the conservation of threatened or endangered species.

58. Describe the history of grizzly bears in the Greater Yellowstone Ecosystem.

59. Identify a benefit for animals after being listed as threatened or endangered.

60. List the animals that have been removed from the endangered species list.

61. Explain how the Endangered Species Act has sparked controversy.

62. What are current challenges for the Endangered Species Act?

63. Define the Convention on International Trade in Endangered Species of Wild Fauna and Flora (CITES).

64. What agency oversees the Red List in the United States?

Conserving Entire Ecosystems

65. Describe the ecosystem approach to conserving biodiversity. Give an example of the ecosystem approach.

66. Figure 59.14: Identify the trend in amount of protected land and ocean from 1960 to 2020.

67. List the questions to consider when protecting land or water.

68. Explain how the theory of island biogeography relates to protected areas.

69. Identify and explain SLOSS.

70. Describe the final consideration regarding size and shape of protected areas.

71. Describe biosphere reserves.

72. List and describe the different zones of a biosphere reserve.

73. Describe the Big Bend National Park in Texas as a biosphere reserve.

74. Explain the concept of restoration ecology.

75. Describe two current restoration ecology projects occurring in the United States.

Visual Representation 9: Threats to biodiversity

76. Complete the following chart by listing the details for each different species:

Species	Conservation Status	Critical Functions in the Ecosystem	Threats to the Species	Current Conservation Efforts
Whitebark pine (*Pinus albicaulis*)				
Mexican long-nose bat (*Leptonycteris nivalis*)				
Devil's Hole pupfish (*Cyprinodon diabolis*)				
Elkhorn coral (*Acropora palmata*)				
African savanna elephant (*Loxodonta Africana*)				
Southern cassowary (*Casuarius casuarius*)				

Pursuing Environmental Solutions

Swapping Debt for Nature

77. Describe three reasons for the expense of preserving biodiversity.

78. Describe the reason developing nations have so much debt with developed countries.

79. Describe Thomas Lovejoy's "debt-for-nature" swap and a possible method for implementation.

80. Explain the Guatemala debt-for-nature swaps.

81. Describe three benefits from the Guatemala debt-for-nature.

82. List the criteria to take part in the swap program.

83. List four countries involved with the swap program.

Science Applied 9

How can we bring back biodiversity?

84. Describe New Zealand biodiversity prior to 700 years ago.

85. Explain how New Zealand biodiversity changed 700 years ago.

86. Describe the initiative announced in 2016 from the New Zealand government.

How can we begin removing invasive predators?

87. Describe Zealandia.

88. Give an example of Zealandia's biodiversity success.

89. Explain the methods that will be used to reach Predator-Free 2050.

Is there opposition to the Predator-Free 2050 plan?

90. Explain the opposition to the Predator-Free 2050.

91. Explain the proponents' argument for the Predator-Free 2050.

92. List non-native species not a part of the eradication program.

How do we move forward?

93. Identify the key to success of the Predator-Free 2050 plan.

94. Identify the economic considerations of the plan.

Read "Do the Math: Estimating Percentages in the Conservation Status of North American Mammals" on page 698. Try "Your Turn." For more math practice, do the following problem. Remember to show your work. Use a separate sheet of paper if necessary.

Complete the table showing the total number and percentages of all North American Birds on the IUCN Red List.

Red List Category	Number of Species	Percent of Species
Extinct	31	
Critically Endangered	25	
Endangered	31	
Vulnerable	54	
Near Threatened	64	
Least Concern	853	
Data Deficient	1	
TOTAL		

Data from: https://www.iucnredlist.org/search

After You Read the Module

Review Key Terms
Match the key terms on the left with the definitions on the right.

_____ 1. Endangered species

_____ 2. Lacey Act

_____ 3. Marine Mammal Protection Act

_____ 4. Endangered Species Act

_____ 5. Convention on International Trade in Endangered Species of Wild Fauna and Flora (CITES)

a. A 1973 treaty formed to control the international trade of threatened plants and animals.

b. Species that are likely to go extinct in the near future.

c. A 1972 U.S. law that prohibits the killing of all marine mammals in the United States and prohibits the import or export of any marine mammal body parts.

d. A U.S. act that prohibits interstate shipping of all illegally harvested plants and animals.

e. A 1973 U.S. law designed to determine which species can be listed as threatened or endangered and prohibits the harming of such species.

UNIT 8 Review Exercises

Check Your Understanding

Review "Learning Goals Revisited" at the end of each module in Unit 1 of your textbook. Compare the notes you took while reading each module. Complete these exercises to review the unit. Use a separate sheet of paper if necessary.

1. Write out the chemical equations of stratospheric ozone formation and the chemical equation for the destruction of ozone from CFCs.

2. Explain the Montreal Protocol.

3. Summarize the greenhouse effect.

4. List the anthropogenic sources of greenhouse gases.

5. Explain how carbon dioxide levels have changed since 1958.

6. Identify the predicted increase in global temperatures by 2100.

7. Describe the following positive feedback loops:

   (a) Melting glaciers and global warming.

   (b) Warming soils and the melting of permafrost leading to further global warning.

8. Explain the difference between coral bleaching and ocean acidification.

9. Identify and explain the acronym used for declining biodiversity. Give an example for each item.

10. Summarize the aims of following laws and treaties.

    *   Lacey Act

    *   CITES

    *   Marine Mammal Protection Act

    *   Endangered Species Act

Practice for Free-Response Questions

Complete this exercise to build and practice the skills you will need to answer free-response questions on the exam. Use a separate sheet of paper if necessary.

1. A volcanic eruption emits a large quantity of ash into the atmosphere. Describe how volcanic eruptions can affect the climate.

2. The contribution of each gas to global warming depends in part on its greenhouse warming potential. The greenhouse warming potential (GWP) of a gas estimates how much a molecule of any compound can contribute to global warming over a period of 100 years relative to one molecule of CO_2, which is defined as having a GWP 1.

    (a) Explain why water vapor has such a low global warming potential.

    (b) Explain why CO_2 is the greatest contributor to greenhouse effect.

3. Charles Keeling observed a decrease in carbon dioxide each spring. Explain the cause of decreased carbon dioxide levels in the spring in the Northern Hemisphere.

Unit 9 Multiple-Choice Review Exam

1. Which greenhouse gas traps the majority of outgoing infrared radiation?
   (a) water vapor
   (b) carbon dioxide
   (c) sulfur dioxide
   (d) nitrogen dioxide

2. Which causes a decline in biodiversity?
   (a) increasing locations of biosphere reserves
   (b) decreasing edge habitats
   (c) overharvesting
   (d) limits on hunting or fishing

3. The greatest cause of biodiversity decline is
   (a) habitat loss.
   (b) alien species.
   (c) overharvesting.
   (d) pollution.

4. Which describes invasive species?
   (a) They spread slowly.
   (b) They have beneficial effects on native species.
   (c) They can be introduced by humans.
   (d) They have a mutualist relationship with native species.

5. Based on those species for which scientist have reliable data, which is most at risk of becoming threatened or near-threatened with extinction?
   (a) birds
   (b) amphibians
   (c) reptiles
   (d) fish

6. The greenhouse effect happens when
   (a) the ozone layer is decreased.
   (b) infrared radiation is absorbed and emitted back to Earth.
   (c) UV light is trapped at Earth's surface.
   (d) CFCs destroy the stratospheric ozone layer.

7. Which has the greatest greenhouse warming potential?
   (a) carbon dioxide
   (b) chlorofluorocarbons
   (c) methane
   (d) nitrous oxide

8. Which greenhouse gas comes from automobiles?
   (a) nitrous oxide
   (b) methane
   (c) sulfur dioxide
   (d) chlorofluorocarbons

9. Which is an anthropogenic cause of greenhouse gas release?
   (a) volcanic eruptions
   (b) use of asbestos
   (c) coal-burning power plants
   (d) evaporation

10. What layer of the Earth's atmosphere absorbs UV-C rays?
    (a) exosphere
    (b) mesosphere
    (c) troposphere
    (d) stratosphere

11. Which produces greenhouse gases and can also increase levels of mercury in the environment?
    (a) refrigerators
    (b) coal
    (c) automobiles
    (d) landfills

12. In response to the decrease in stratospheric ozone, the _____ helped reduce CFC production.
    (a) CITES
    (b) Kyoto Protocol
    (c) Paris Climate Agreement
    (d) Montreal Protocol

13. Why do levels of carbon dioxide in Earth's atmosphere vary seasonally?
    (a) More fossil fuels are burned in the summer for heat.
    (b) Livestock production increases each spring.
    (c) Photosynthesis varies each season.
    (d) Landfills increase in size as more trash is produced during summer months.

14. Which is a likely effect of warmer temperatures on the environment?
    (a) melting polar ice caps
    (b) altered biogeochemical cycles
    (c) increased cold spells
    (d) larger hole in the stratospheric ozone layer.

15. Which agreement aims to control the emissions that contribute to global warming?
    (a) Montreal Protocol
    (b) Kyoto Protocol
    (c) Endangered Species Act
    (d) Clean Air Act

16. _____ refers to storing carbon in agricultural soils to return atmospheric carbon to longer-term storage in the form of plant biomass.
    (a) Agricultural restoration
    (b) Soil reclamation
    (c) Carbon sequestration
    (d) Carbon dioxide scrubber

17. Which location in a biosphere reserve has the least human activity?
    (a) buffer zone
    (b) transition area
    (c) core zone
    (d) settlement area

18. Ocean acidification occurs
    (a) when stressed corals eject their algae.
    (b) when pH levels of the ocean fall.
    (c) after coral bleaching.
    (d) when pH levels of the ocean rise.

19. In the Paris Climate Agreement 196 countries pledged to
    (a) reduce coral bleaching.
    (b) stop the development of CFCs.
    (c) reduce greenhouse gases by 5.2 percent by 2012.
    (d) keep global warming less than 2°C.

20. Oceans have absorbed about _____ percent of the excess heat produced through global warming.
    (a) 75
    (b) 50
    (c) 90
    (d) 10

21. Chlorofluorocarbons can be found in
    (a) coal.
    (b) ozone.
    (c) UV-C.
    (d) refrigerants.

22. The country with the greatest carbon dioxide emission is
    (a) Canada.
    (b) India.
    (c) United States.
    (d) China.

23. The greenhouse gases that began to rise dramatically after the 1800s are:
    (a) carbon dioxide, methane, and nitrous oxide
    (b) carbon monoxide and methane
    (c) methane and nitrous oxide
    (d) carbon monoxide and nitrogen

24. The polar ice cap near the North Pole has decreased by _____ percent per decade.
    (a) 10
    (b) 13
    (c) 17
    (d) 23

25. Melting permafrost will lead to
    (a) overlying lakes becoming smaller.
    (b) the release of less methane.
    (c) less decomposition.
    (d) no change for overlying lakes.

26. The pied flycatcher of the Netherlands
    (a) has been provided a new food source due to global warming.
    (b) has a lower food choice due to global warming.
    (c) has no change in a food source due to global warming.
    (d) has fewer nesting sites because of global warming.

27. Effects of ocean warming in the North Sea and fish distribution through the years can be summarized as:
    (a) bottom waters have cooled and the fish populations decreased.
    (b) bottom waters have warmed and the fish populations have decreased.
    (c) bottom waters have cooled and the fish populations increased.
    (d) bottom waters have warmed and the fish populations have increased.

28. In response to the elevated pH of ocean water
    (a) the pencil urchin decreased in size.
    (b) the pencil urchin increased in size.
    (c) the pencil urchin stayed the same size.
    (d) the pencil urchin had thicker shells.

29. The kudzu vine is an example of
    (a) an endangered species.
    (b) a native species.
    (c) an invasive species.
    (d) a domesticated species.

30. The greatest decline and extinction of species is due to
    (a) pollution.
    (b) invasive species.
    (c) climate change.
    (d) habitat destruction.

Full-Length Practice Exam 1

This full-length practice exam contains two sections. Section I consists of 80 multiple-choice questions and Section II consists of 3 free-response questions.

You will have 90 minutes to complete the multiple-choice section of the exam. As you will not be penalized for incorrect answers, you should answer every question on the test. If you do not know an answer to a question, try to eliminate any incorrect answer choices and take your best guess. Do not spend too much time on any one question. If you know the question is going to take a while to answer, you should skip it and come back to it at the end.

You will have 70 minutes to complete the free-response section of the exam. Be sure to answer each part of the question and to provide thorough explanations using the terms and themes you have learned in the course. Also be sure to show your work whenever you use math to solve a problem.

Graphing calculators are allowed on both sections of the exam.

SECTION I: Multiple-Choice

Choose the best answer for questions 1-80.

1. Which of the following characteristics is expected in a country going through a demographic transition?
    (a) Birth rates rise.
    (b) Death rates rise.
    (c) Birth rates fall.
    (d) Death rates remain high.

2. Which of the following is associated with electricity generation through solar power?
    (a) installing a photovoltaic cell on the roof
    (b) hanging blinds in a window
    (c) planting a deciduous tree outside a west facing window
    (d) installing a living roof

3. Which is the correct order of soil particles from largest to smallest?
    (a) sand, clay, silt
    (b) silt, clay, sand
    (c) sand, silt, clay
    (d) clay, silt, sand

4. A scientist performs an experiment with brine shrimp. The scientist adds increasing dosages of a particular pesticide to populations of brine shrimp and determines that a dosage of 10 ml/L of the pesticide kills half of the shrimp in the sample. What has the scientist discovered in the experiment?
    (a) the LD-50
    (b) the effective dose
    (c) the toxicity level
    (d) the ED-50

5. Which is an example of an indicator species for an old growth forest?
    (a) elephant
    (b) spotted owl
    (c) leopard frog
    (d) box turtle

6. Radioactive waste generated in the United States is currently being stored
    (a) at Yucca Mountain.
    (b) in the ocean.
    (c) in Mexico.
    (d) at the nuclear plants where the waste is produced.

7. Which is a concern about excessive use of fertilizer in yards and gardens?
    (a) It will lead to an increased number of pests.
    (b) It will contribute to eutrophication in nearby waterways.
    (c) It will hasten the extinction of endangered species.
    (d) It will damage the niche of a keystone species.

8. Which of the following greenhouse gases cycles too quickly in the atmosphere to significantly contribute to global warming?
   (a) methane
   (b) CFCs
   (c) carbon dioxide
   (d) water vapor

9. Alum is used in primary treatment of wastewater. Which of the following statements describes how alum is used in sewage treatment plants?
   (a) Alum screens out the sewage.
   (b) Alum decontaminates the sewage.
   (c) Alum allows components of the sewage to clump and sink.
   (d) Alum aerates the sewage.

10. Which of the following might explain why a country has a replacement level fertility of 3.5?
    (a) Women are working outside the home.
    (b) The life span is only 45 years.
    (c) The country has a high infant mortality rate.
    (d) The country has a large population of women above childbearing age.

11. Which of the following statements correctly describes the efficiency of a typical coal-fired electricity plant?
    (a) The efficiency is 100 percent because coal has the highest energy conversion rate.
    (b) The efficiency is 75 percent because some of the energy is lost as light and sound.
    (c) The efficiency is 50 percent because half of the energy is lost as pollution.
    (d) The efficiency is 30 percent because much of the heat is lost to the environment.

12. Habitat loss, invasive species, pollution, overpopulation, climate change, and overharvesting have contributed to which of the following?
    (a) increasing numbers of endangered species
    (b) mountain top removal for coal mining
    (c) cultural eutrophication in surface water ways
    (d) ozone depletion in the stratosphere

13. A city with a population of 10,000 has 50 births, 30 deaths, 30 immigrants and 10 emigrants in a year. What is the net annual percentage growth rate?
    (a) 120 percent
    (b) 40 percent
    (c) 8 percent
    (d) 0.4 percent

14. The Earth is tilted approximately 23 degrees. This is most closely associated with
    (a) the changing of seasons.
    (b) the high and low tides.
    (c) the solar lights.
    (d) the Coriolis effect.

15. Radon is found in bedrock and can seep into homes through basements and other means. Which statement describes why radon is a concern?
    (a) Radon causes asthma.
    (b) Radon causes brain damage in children.
    (c) Radon damages foundations.
    (d) Radon causes lung cancer.

16. Which of the following describes an outcome of using nuclear power?
    (a) Radioactive waste with long half-lives is generated and requires disposal.
    (b) Surface waters used to cool the reactor core become radioactive.
    (c) Significant CO_2 emissions are generated.
    (d) Birth rates decrease for the areas adjacent to the nuclear power plant.

17. Which of the following describes a disadvantage of producing genetically modified crops in agriculture?
    (a) The genetically modified crops are pest resistant.
    (b) The genetically modified crops have improved nutrition.
    (c) The genetically modified crops have a greater yield.
    (d) The genetically modified crops can decrease the biodiversity of crops.

18. Which is a nonrenewable energy source?
    (a) nuclear
    (b) solar
    (c) geothermal
    (d) biomass

Use the graph below to answer questions 19-20.

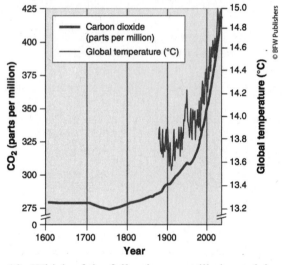

19. Which of the following most likely explains the trend for carbon dioxide levels over the past 100 years?
    (a) increased production of chlorofluorocarbons
    (b) decreased reliance on meat-based diets
    (c) increased combustion of fossil fuels
    (d) increased use of nuclear power

20. What environmental impact is most likely associated with the change in temperature on the graph?
    (a) increased number of specialist species
    (b) decreased size of Arctic ice sheets
    (c) increased seismic activity
    (d) decreased insect-carried diseases

Use the following graph to answer question 21.

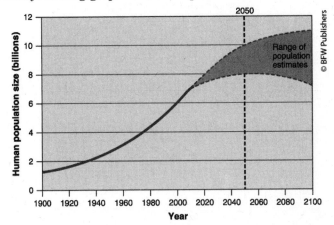

21. According to the graph, from 1800 to 2010 the human population displayed which of the following?
   (a) linear growth.
   (b) bell-shaped growth.
   (c) exponential growth.
   (d) a sharp decline.

22. Consider two different communities each with 100 organisms.
   Community 1 has 10 different species with 10 individuals in each species.
   Community 2 has 5 different species with the following numbers of individuals:
   - species 1 = 40
   - species 2 = 20
   - species 3 = 20
   - species 4 = 10
   - species 5 = 10

   Which of the following describes a difference between the two communities?
   (a) Community 1 has more species richness than community 2.
   (b) Community 2 has more species richness than community 1.
   (c) Community 2 has more species evenness than community 1.
   (d) Community 1 is more stable than community 2.

Use the following diagram to answer questions 23 and 24.

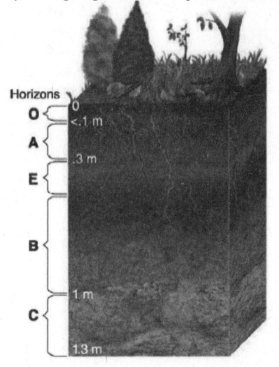

23. Which layer consists of organic detritus?
    (a) O
    (b) A
    (c) B
    (d) E

24. Which layer is known as the subsoil and has very little organic matter?
    (a) O
    (b) A
    (c) B
    (d) E

25. If a population is growing at a rate of 7 percent per year, how many years will it take for the population to double?
    (a) 10 years
    (b) 17 years
    (c) 25 years
    (d) 70 years

Read the following excerpt and answer questions 26-28 based on the article.

Reversing Human Impacts on a Salty Lake

Located between the deserts of the Great Basin and the mountains of the Sierra Nevada, California's Mono Lake is an unusual site. It is characterized by eerie towers of limestone rock known as tufa, glassy waters, unique animal species, and frequent dust storms. Mono Lake is a terminal lake, which means that water flows into it but does not flow out. As water moves through the mountains and desert soil, it picks up salt and other minerals, which it deposits in the lake. As the water evaporates, these minerals are left behind. Over time, evaporation has caused a buildup of salt concentrations so high that the lake became as salty as the ocean. While the lake is too salty for fish and many other lake species to survive, there are a few species that are specially adapted to live there, including the Mono brine shrimp (*Artemia monica*) and larvae of the Mono Lake alkali fly (*Ephydra hians*). The shrimp and the fly larvae consume microscopic algae, millions of tons of which grow in the lake each year. In turn, large flocks of migrating birds, such as sandpipers, gulls, and flycatchers, use the lake as a stopover, feeding on the brine shrimp and fly larvae. The lake is an oasis on the migration route for these birds and they have come to depend on its food and water resources. The health of Mono Lake is therefore critical for many species.

In 1941, the City of Los Angeles built an aqueduct to divert water from the streams feeding Mono Lake so it could be used by residents of the city. With less stream water feeding the lake, the surface of the lake dropped 14 m (45 feet) by 1982. The lake now had half as much water and the salinity of the water doubled to more than twice that of the ocean. The salt killed algae in the lake; without the algae to eat, the brine shrimp also died. Most migrating birds stayed away, and newly exposed land bridges allowed coyotes from the desert to prey on the colonies of nesting birds that remained. These changes underscored the interconnectedness of the Mono Lake food web.

26. Which of the following best describes the author's claim in the article?
    (a) Human activities can significantly alter natural ecosystems.
    (b) Terminal lakes cannot support stable food webs.
    (c) Water diversion projects must be reversed to prevent evaporation from lakes.
    (d) Aquatic ecosystems cannot be effectively converted to terrestrial ecosystems.

27. Based on the passage, which of the following best describes the alkali fly?
    (a) primary producer
    (b) primary consumer
    (c) secondary consumer
    (d) top predator

28. Which of the following would best reduce the environmental impacts at Mono Lake?
    (a) removing the migrating bird nests to prevent coyotes from having sufficient food sources
    (b) removing significant amounts of dry salt from the exposed surfaces at the lake
    (c) using groundwater to refill Mono Lake to its original level
    (d) reversing the water diversion so that Mono Lake has an increased input

29. The discovery of coliform bacteria in a river would most likely indicate which of the following?
    (a) Chemical fertilizers used by nearby homeowners have run off into the river.
    (b) An invasive species is present in the river.
    (c) Sewage from a wastewater treatment plant has leaked into the river.
    (d) A factory is dumping industrial waste into the river.

30. Which of the following age structure diagrams identifies a country experiencing the fastest growth rate?

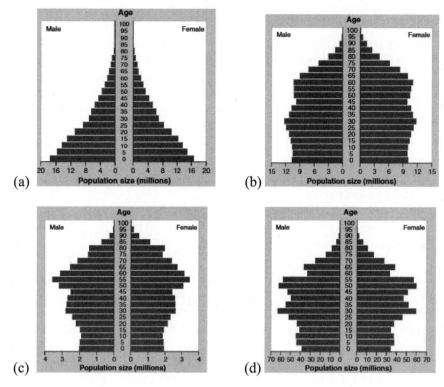

(a)

(b)

(c)

(d)

31. Which trait is most likely associated with a niche specialist?
    (a) has a highly varied diet
    (b) can colonize new areas rapidly
    (c) is engaged in mutualism
    (d) is especially susceptible to environmental change

32. Which of the following best prevents soil erosion?
    (a) no till plowing
    (b) overgrazing
    (c) logging
    (d) deforestation

33. The order of coal from the lowest energy content to highest energy content is
    (a) anthracite, lignite, sub-bituminous, bituminous.
    (b) anthracite, sub-bituminous, bituminous, lignite.
    (c) sub-bituminous, bituminous, lignite, anthracite.
    (d) lignite, sub-bituminous, bituminous, anthracite.

34. Which fossil fuel has the largest, most accessible supply on Earth?
    (a) oil
    (b) natural gas
    (c) biomass
    (d) coal

35. Which of the following is a valid environmental concern about concentrated animal feeding operations (CAFOs)?
    (a) They require more land than free range farms.
    (b) They cause most of the overgrazing seen in the central United States.
    (c) They increase the cost of meat at the market.
    (d) They increase the potential for groundwater contamination from animal waste.

Use the following figure to answer questions 36-38.

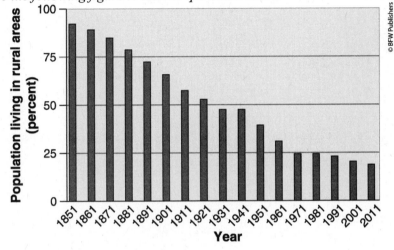

36. Which statement is supported by the information in the graph?
    (a) The majority of the population has been living in urban areas since the 1930s.
    (b) The population size has been declining over the past 150 years.
    (c) Decreased transportation opportunities require more people to leave rural areas.
    (d) New farming practices required people to move out of rural areas after 1930.

37. Which of the following most likely occurred after 1950 based on the information in the graph?
    (a) decreased reliance on petroleum for transportation
    (b) increased urban sprawl
    (c) increased migration of people to rural areas
    (d) decreased transportation needs

38. If the United States population was 230 million people in 1981, approximately how many people were living in rural areas in that year?
    (a) 25 million
    (b) 57.5 million
    (c) 115 million
    (d) 172.5 million

Examine the graph for worldwide annual energy consumption to answer questions 39 and 40.

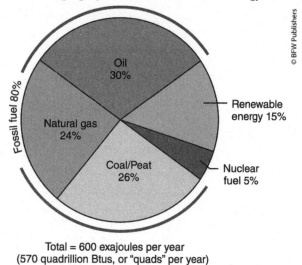

Total = 600 exajoules per year
(570 quadrillion Btus, or "quads" per year)

39. How many exajoules per year are produced from renewable energy resources?
  (a) 15 exajoules
  (b) 90 exajoules
  (c) 150 exajoules
  (d) 600 exajoules

40. Which statement correctly describes how the graph would change if it only represented a less developed nation?
  (a) Nuclear energy and petroleum percentages would decrease.
  (b) Coal and nuclear energy percentages would increase.
  (c) There would be no renewable energy represented in the graph.
  (d) All of the energy would come from oil and natural gas.

Use the graph showing global growth of wind energy capacity to answer questions 41-42.

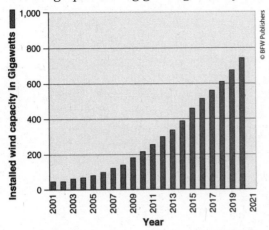

41. According to the graph, which statement describes the changes in wind energy capacity from 2001 to 2019?
   (a) A decrease in petroleum reserves required the installation of new wind turbines.
   (b) An increase in wind turbine installation increased the total energy generated by renewable resources.
   (c) The failure of nuclear power plants required an increase in the number of hours each turbine is used to generate electricity.
   (d) Stronger winds caused by increasing global temperatures generated more electricity from wind power.

42. Which of the following represents the percent change in wind energy capacity from 2009 to 2014?
   (a) 50%
   (b) 100%
   (c) 200%
   (d) 400%

Use the figure below to answer question 43.

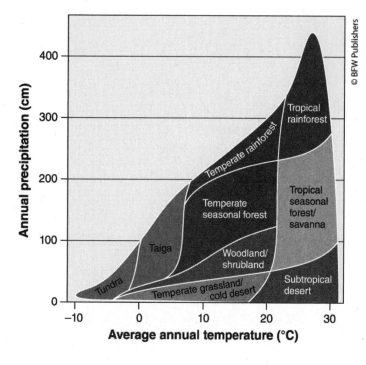

43. Which biome most likely has permafrost?
    (a) tundra
    (b) grassland
    (c) tropical rainforest
    (d) temperate rainforest

44. After several years of using a particular pesticide, a farmer notices that it is killing fewer insect pests. Which of the following statements best explains this phenomenon?
    (a) The insect pests have become resistant to the pesticide.
    (b) The pesticide has leeched essential nutrients from the field.
    (c) Other conditions have become more favorable for the insect pests.
    (d) The insects have migrated to other croplands.

Read the following excerpt and use the information to answer questions 45-48.

In the state of Washington stands Mount St. Helens. Mount St. Helens is a large mountain that was formed by volcanoes, along with the other mountains in the Cascade Range, which had remained quiet for more than a hundred years. All of this changed on May 18, 1980…At 8:32 AM Mount St. Helens erupted, blowing away the north face of the mountain. It is estimated that the mountain released 540 million tons of ash, which would fill a football field with ash stacked 150 miles high. Although many volcanoes also release lava, this eruption did not have flows of lava coming down the mountain…

…The explosion of heat, rocks, and ash scoured and scorched a massive area around the mountain. It was expected that it would take decades to centuries for the area to undergo primary succession. Interestingly, the large field of ash did not have to wait for plant seeds to arrive to initiate ecological succession. Researchers discovered that a few plants were still alive under the snow and ash layer and these plants were able to poke up through the layers and jump-start the process of succession. As a result, some climax species were present early in the succession process while some early succession species did not show up until much later. In addition, the scarce surviving plants attracted small mammals and birds, which likely carried additional seeds into the area in their digestive systems. Researchers also discovered that pocket gophers, which were in their underground burrows at the time of the eruption, below a layer of snow and soil, had survived and were able to dig out of the volcanic materials. In doing so, they mixed the volcanic ash, which contained no organic matter, with the underlying soil; this made the ground more hospitable for plants to grow. They also learned that it was better to leave the dead trees lying on the ground rather than remove them, because the trees helped prevent erosion of the soil. Today, the once destroyed forests are a thicket of willows, alders, huckleberries, and young fir trees.

In 1980, volcano research was still in its infancy. Over the past 40 years, however, Mount St. Helens has become a living laboratory for hundreds of environmental scientists who are working to understand the process that cause volcanoes around the world, to predict when they will erupt, and to forecast how the processes of ecological succession will subsequently unfold.

45. Which of the following best identifies the author's claim in the reading?
    (a) Primary succession always takes longer after volcanic eruption.
    (b) Natural disruptions to ecosystems offer opportunities to study ecological changes.
    (c) Changes to ecosystems from natural disasters are permanent.
    (d) Natural disruptions to ecosystems are more severe than human impacts.

46. Which of the following pieces of evidence would support that biodiversity was greater forty years after the volcanic eruption?
    (a) More seeds germinate from the soil/ash mixture.
    (b) Average tree height is taller forty years after the eruption.
    (c) A larger variety of species is present in the area affected by the volcano.
    (d) There are fewer predators in the food web of the affected area.

47. Based on the information from the excerpt, what role did the pocket gophers play in the recovering ecosystem?
    (a) primary producer
    (b) indicator species
    (c) top predator
    (d) keystone species

48. Ecologists recommended that the trees destroyed by the volcanic eruption be left on the ground after the explosion. Which reason would the ecologists have for making this recommendation?
    (a) Fallen trees can be used for timber and lumber.
    (b) Fallen trees can reduce erosion of volcanic ash into local waterways.
    (c) Fallen trees can prevent other seismic activity in the area following the eruption.
    (d) Fallen trees can prevent invasive species from colonizing the area.

Use the following figure to answer question 49.

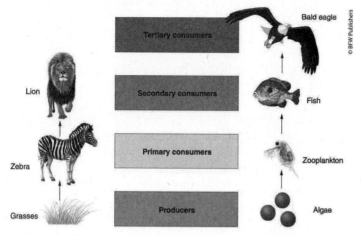

(a) Terrestrial food chain   **(b)** Aquatic food chain

49. Based on the terrestrial food chain, what is the predicted energy availability to lions if there are 250,000 kJ of energy at the grasses?
    (a) 25 kJ.
    (b) 250 kJ
    (c) 2500 kJ
    (d) 25,000 kJ

50. A forest ecosystem has an NPP of 3.05 kg $C/m^2/year$ and a GPP of 4.5 kg $C/m^2/year$. How much carbon is lost to respiration?
    (a) 1.45 kg $C/m^2/year$
    (b) 7.55 kg/$C/m^2/year$
    (c) -1.45 kg $C/m^2/year$
    (d) 13.73 kg $C/m^2/year$

Use the following figure to answer question 51.

51. Which environmental phenomena is demonstrated in the model?
    (a) ENSO
    (b) Coriolis effect
    (c) Rain shadow effect
    (d) Albedo from light-colored surfaces

Use the following figure to answer questions 52-54.

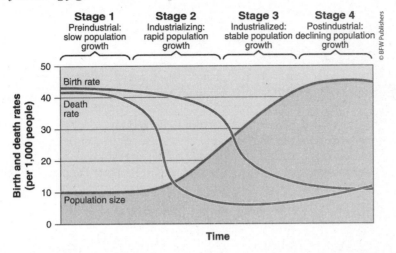

52. In which stage do the death rates fall the fastest as a country moves through the stages of demographic transition?
    (a) Stage 1
    (b) Stage 2
    (c) Stage 3
    (d) Stage 4

53. Which statement explains the population stabilization in Stage 4?
    (a) Humans have reached their carrying capacity and the population cannot grow.
    (b) Humans have migrated to less developed areas to exploit more natural resources.
    (c) The population has moved from rural to urban areas, expanding the cities in size and population.
    (d) People have fewer children since fewer offspring die in infancy and women are more likely to work outside the home.

54. Improved sanitation and nutrition, but a continued dependence on an agricultural society is likely found in which stage of the Demographic Transition Model?
    (a) Stage 1
    (b) Stage 2
    (c) Stage 3
    (d) Stage 4

Use the following diagram to answer questions 55 and 56.

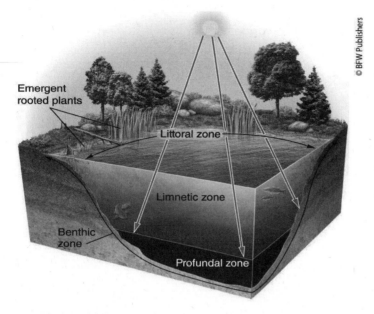

55. In which zone of the lake would you expect to find the smallest biomass of photosynthetic algae?
    (a) littoral
    (b) limnetic
    (c) profundal
    (d) benthic

56. In which zone of the lake would you find algae but no rooted plants?
    (a) littoral
    (b) limnetic
    (c) profundal
    (d) benthic

57. Which of the following predicts how deforestation of the riparian zone would affect a river system?
    (a) It would increase sediment levels.
    (b) It would increase oxygen levels.
    (c) It would increase the presence of invasive species.
    (d) It would increase the number of fish present.

58. Which of the following describes why ozone levels peak after nitrogen oxide levels on a summer day?
    (a) Low levels of nitrogen oxides cause high levels of ozone.
    (b) Nitrogen oxides are released from natural sources while ozone is released from anthropogenic sources.
    (c) An increase in sunlight can increase ozone production from NOx in the atmosphere.
    (d) Nitrogen oxides are removed from the atmosphere by ozone.

59. Which statement describes the effect of an increase in ozone in the troposphere?
   (a) Ozone acts as a respiratory irritant in humans.
   (b) Ozone decreases the pH of surface waters.
   (c) Ozone blocks sunlight from reaching the Earth's surface.
   (d) Ozone increases rates of skin cancer and cataracts.

60. The Himalayan Mountains, one of the youngest mountain chains in the world, was likely made at which of the following geologic locations?
   (a) a transform plate boundary
   (b) a divergent plate boundary
   (c) a convergent plate boundary
   (d) a hotspot

Use the following diagram to answer questions 61 and 62.

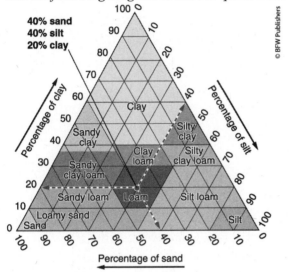

61. A farmer is observing a lack of drainage in the fields after heavy rains. Which soil profile is most likely to be found in a sample of the farmer's soil?
   (a) 80% clay- 10% silt, 10% sand
   (b) 20% clay- 50% silt, 30% sand
   (c) 10% clay- 10% silt, 80% sand
   (d) 30% clay- 10% silt, 60% sand

62. A soil sample with a soil profile of 20 percent sand, 30 percent silt, 50 percent clay would be classified as which soil type?
   (a) sand
   (b) silt
   (c) clay
   (d) loam

63. Which of the following describes leachate?
   (a) the waste products from invertebrates that live in the soil
   (b) the toxic residue that remains after incinerating MSW
   (c) the liquid that contains pollutants from MSW or contaminated soil
   (d) the radioactive waste from a nuclear plant

64. Which of the following materials could be reduced from the waste stream by either composting or recycling to avoid taking up space in a typical landfill?
    (a) aluminum
    (b) plastic
    (c) glass
    (d) paper

65. Farmers can use biological controls to reduce pesticides used on crops. However, sometimes biological controls can
    (a) overpopulate an area.
    (b) cause the pest to multiply.
    (c) become resistant.
    (d) target both beneficial and pest species.

66. Many farmers have found that over a number of years they must spray more and more pesticide on their crops for the same result. This is known as
    (a) insect migration.
    (b) the pesticide treadmill.
    (c) GMOs.
    (d) speciation.

67. Which law enforces the cleanup of hazardous waste sites?
    (a) CERCLA (Superfund)
    (b) Clean Air Act
    (c) RCRA
    (d) Clean Water Act

68. Which might explain the loss of significant numbers of lichen in an ecosystem?
    (a) The ecosystem has an invasive species.
    (b) The ecosystem is being affected by acid deposition.
    (c) The ecosystem is being affected by thermal pollution.
    (d) The ecosystem is being affected by eutrophication.

69. Which chemicals and sources are responsible for acid deposition?
    (a) carbon dioxide released from combustion of biomass
    (b) mercury released from combustion of coal
    (c) sulfur dioxide released from fossil fuel combustion
    (d) CFCs used as refrigerants and coolants in old appliances

70. Which is a characteristic of an *r*-selected species?
    (a) small size
    (b) specialist
    (c) at risk for endangerment
    (d) few offspring

71. Sulfur dioxide is a component of industrial smog. Which pollution control measure is most effective for sulfur dioxide?
    (a) catalytic converter
    (b) scrubber
    (c) electrostatic precipitator
    (d) bag house filter

72. Which of the following is an environmental problem associated with sediment pollution in surface waters?
    (a) It prevents the mobility of fish and aquatic invertebrates.
    (b) It increases turbidity in the water which causes a reduction in productivity of plants.
    (c) It increases the effects of thermal pollution.
    (d) It cools the surface water, reducing the metabolism of aquatic species.

73. Because the U.S. population has a TFR of 1.9 and high net migration, it may be best described as
    (a) a country experiencing population momentum.
    (b) a country with declining population growth.
    (c) a country with stable population growth.
    (d) a country with rapid population growth.

74. Which international agreement effectively banned the use of CFCs?
    (a) Montreal Protocol
    (b) Kyoto Protocol
    (c) CITIES
    (d) The Clean Air Act

Use the figure below to answer questions 75-77.

75. Which concept is best illustrated in the figure?
    (a) Tragedy of the commons
    (b) biomagnification
    (c) groundwater contamination from leachate
    (d) eutrophication

76. How much more concentrated is the persistent organic pollutant in the threespine stickleback compared to the plankton?
    (a) 1.04 times more concentrated
    (b) 0.22 times more concentrated
    (c) 6.5 times more concentrated
    (d) 650 times more concentrated

77. Which would be an example of a persistent organic pollutant shown in the model?
    (a) Radon
    (b) Phosphate
    (c) Nitrate
    (d) DDT

78. What is the doubling time of the United States if the growth rate is 0.7 percent?
    (a) 10 years
    (b) 70 years
    (c) 100 years
    (d) 700 years

79. The Kyoto Protocol addresses which environmental concern?
    (a) habitat loss
    (b) biodiversity loss
    (c) marine mammals being killed
    (d) climate change

80. Which of the following correctly identifies an indoor air pollutant and its source?
    (a) radon originates from paints and solvents
    (b) particulate matter originates from combustion of biomass
    (c) VOCs originate from the bedrock
    (d) carbon monoxide originates from passive solar systems

SECTION II: Free-Response Questions

Write your answer to each part clearly. Support your answers with relevant information and examples. Where calculations are required, show your work. Use a separate piece of paper if necessary.

1. An experiment was performed to determine the effect of snail density on species richness of algae in aquatic ecosystems. The data from the experiment is shown below.

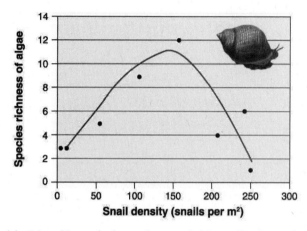

   (a) **Identify** an independent variable and a dependent variable from the experiment.

   (b) **Identify** the snail density that resulted in the highest species richness for algae.

   (c) **Describe** the trend observed in the results from the experiment. **Explain** why the population density of snails affects the species richness of algae in terms of the trophic levels.

   (d) **Explain** why an increase in species richness of algae increases the productivity of an ecosystem. **Identify** another measure of biodiversity besides species richness for the algae.

   (e) Humans have increased the use of pesticides that have caused an influx of the chemical into the local waterways. **Predict** one impact the pesticides may have on the snails. **Describe** one practice that humans can use to reduce the impacts of pesticides. **Explain** why humans may be less likely to use the practice you described previously.

2. Examine the net productivity for the ecosystems shown below.

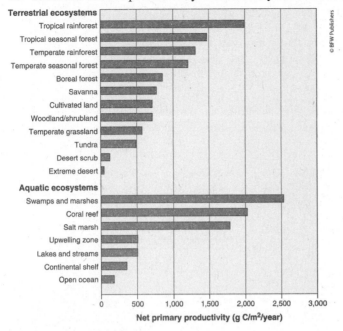

Scientists have been concerned about the effects of the depletion of the stratospheric ozone layer on the health of humans and ecosystems.

(a) **Identify** the terrestrial and aquatic ecosystems that have the highest net productivity.

(b) **Describe** one limiting factor for terrestrial ecosystems and one limiting factor for aquatic ecosystems that causes the productivity to be low.

(c) **Identify** one regulating ecosystem service and one provisioning ecosystem service provided by swamps and marshes.

(d) There has been an observed decrease in the pH of the ocean over the past 50 years.

    i.   **Describe** one environmental impact to a decreasing ocean pH on the coral reef ecosystems.

    ii.   **Provide** one solution for the environmental impact identified in part i.

(e) **Explain** why ocean levels have been rising over the past century and describe one mitigation that coastal communities can use to prevent impacts from the rise in ocean levels.

3. The Miller family has decided that due to the severe drought in their state, they are going to implement water conservation methods in their home. Currently, the family uses approximately 8,000 gallons of water a month for their family of four and they pay $3.45 per 100 cubic feet of water. There are 748 gallons in 100 cubic feet.

    (a) **Calculate** the family's water bill for one month and show your work.

    (b) The family installs water-saving shower fixtures in both bathrooms, which cuts the amount of water for showering in half. Each member of the family takes one shower a day and the average time spent in the shower is 10 minutes. The old shower fixture used 42 gallons every 10 minutes.

        i. **Calculate** how many gallons will be saved by the family per week with the water-saving fixtures. Show your work.

        ii. **Calculate** the monetary savings from the installation of the water-saving fixtures. Show your work.

    (c) Other than the shower fixtures, the Miller family wants to install other water-saving devices in their home. **Describe** two ways the family could conserve more water in their home.

    (d) The family also wants to incorporate water-saving methods outside of their home. **Describe** two ways they could accomplish this goal.

Full-Length Practice Exam 2

This full-length practice exam contains two sections. Section I consists of 80 multiple-choice questions and Section II consists of 3 free-response questions.

You will have 90 minutes to complete the multiple-choice section of the exam. As you will not be penalized for incorrect answers, you should answer every question on the test. If you do not know an answer to a question, try to eliminate any incorrect answer choices and take your best guess. Do not spend too much time on any one question. If you know the question is going to take a while to answer, you should skip it and come back to it at the end.

You will have 70 minutes to complete the free-response section of the exam. Be sure to answer each part of the question and to provide thorough explanations using the terms and themes you have learned in the course. Also be sure to show your work whenever you use math to solve a problem.

Graphing calculators are allowed on both sections of the exam.

SECTION I: Multiple-Choice

Choose the best answer for questions 1-80.

1. Which of the following describes a practice of integrated pest management?
   (a) increasing the dosage of pesticide so more insects will be killed
   (b) combining herbicides and pesticides together so that only one application must be delivered
   (c) using pheromones of a predator species to deter pests from attacking a crop
   (d) adjusting the pesticide treatment annually to combat the pesticide treadmill effect

2. Which type of irrigation technique reduces evaporative water loss and involves digging trenches to fill with water?
   (a) flood irrigation
   (b) furrow irrigation
   (c) drip irrigation
   (d) hydroponic irrigation

3. If the average person in the United States uses 1,000 watts of electricity 24 hours a day for 365 days per year, how many kW of energy does the average person use in a year?
   (a) 1,000 kW
   (b) 3,650 kW
   (c) 3,650,000 kW
   (d) 8,760 kW

Read the excerpt below and answer question 4-6, which follow.

Should Corn Become Fuel?

Corn-based ethanol is big business — so big, in fact, that in an effort to offset demand for petroleum, U.S. policy has resulted in the production of 53 billion liters (14 billion gallons) in 2020. The United States is the largest producer of corn ethanol in the world. Ethanol proponents maintain that substituting ethanol for gasoline decreases air pollution, greenhouse gas emissions, and our dependence on foreign oil. Opponents counter that when we consider all the inputs used to grow and process corn into ethanol, it increases air pollution and greenhouse gas emissions. Moreover, opponents maintain that growing corn and converting it into ethanol uses more energy than we obtain when we burn ethanol for fuel and that the impact of ethanol on reducing our import of foreign oil is very small.

4. Which of the following best identifies the author's claim?
    (a) Corn-based ethanol has environmental benefits but may require more energy than it produces.
    (b) Corn-based ethanol is an economic alternative to expensive fossil fuels.
    (c) Corn-based ethanol has many more years of research before it is ready to implement as a fuel source.
    (d) Corn-based ethanol will not be a fuel resource used commonly in the future.

5. Which of the following describes an environmental benefit of corn-based ethanol compared to traditional gasoline?
    (a) Its combustion does not release carbon dioxide.
    (b) It does not require an energy-input to generate.
    (c) Its combustion releases fewer air pollutants than gasoline.
    (d) Its combustion releases carbon monoxide rather than methane.

6. Which of the following describes a challenge to using corn-based ethanol as a fuel?
    (a) The land used to grow corn for ethanol could be used to grow food for human consumption.
    (b) The land used to grow corn for ethanol is located in foreign countries.
    (c) Corn-based ethanol must be imported from other countries.
    (d) Corn-based ethanol must be taxed more heavily than gasoline.

7. Which of the following processes occurs in the photic zone but not in the abyssal zone of an aquatic ecosystem?
    (a) cellular respiration
    (b) decomposition
    (c) photosynthesis
    (d) upwelling

Use the following figure to answer question 8.

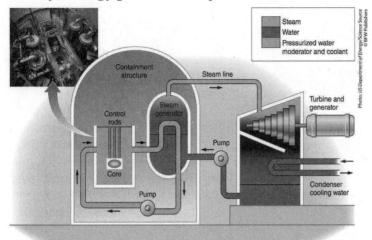

8. Which part of a nuclear reactor is most associated with preventing radiation leaks?
   (a) containment structure
   (b) steam generator
   (c) control rods
   (d) cooling tower

9. The estimate of the average number of children that each woman in a population will bear throughout her childbearing years is
   (a) the total fertility rate.
   (b) average life expectancy.
   (c) the crude birth rate.
   (d) family planning.

10. Which food production activity has the lowest ecological footprint?
    (a) hunting and gathering
    (b) raising grass-fed beef
    (c) far-offshore fishing
    (d) locally produced food sold at a farmer's market

11. Rain that is slightly acidic is responsible for which of the following?
    (a) chemical weathering.
    (b) physical weathering.
    (c) convection.
    (d) subduction.

12. Which of the following describes how deforestation causes changes to the hydrologic cycle?
    (a) There will be increased transpiration and decreased precipitation.
    (b) There will be decreased transpiration and increased evaporation.
    (c) There will be increased precipitation and reduced evaporation.
    (d) There will be reduced evaporation and increased infiltration.

Use the food web below to answer questions 13-15.

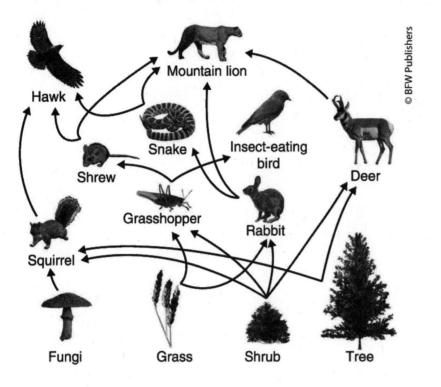

13. At which trophic level does the rabbit feed?
    (a) primary producer
    (b) primary consumer
    (c) secondary consumer
    (d) tertiary consumer

14. Which organism in the food web also contributes to the recycling of organic matter back into the soil?
    (a) Fungi
    (b) Grass
    (c) Shrub
    (d) Tree

15. Which two species compete for the same food source according to the food web?
    (a) Fungi and grass
    (b) Hawk and insect-eating bird
    (c) Grasshopper and rabbit
    (d) Deer and mountain lion

16. A photovoltaic cell is used to
    (a) turn the sun's energy into electricity.
    (b) burn biomass fuel.
    (c) generate wind power.
    (d) generate electricity behind a dam.

17. The pesticide treadmill occurs when
    (a) bioaccumulation in predator species becomes pervasive.
    (b) poor weather conditions make crops more vulnerable to pests.
    (c) pests begin to reproduce at a faster rate.
    (d) a farmer must switch pesticides because resistance develops.

18. Which would cause the greatest decline in biodiversity?
    (a) selective cutting
    (b) clear cutting
    (c) shelter-wood harvesting
    (d) seed-tree harvesting

19. Which is an environmental concern regarding the use of wind power?
    (a) the amount of air pollution generated
    (b) resource depletion
    (c) dependence on the amount of sunlight
    (d) death of birds and bats that collide with turbines

20. Which biotic component of an ecosystem would limit the carrying capacity of primary consumers during a severe drought?
    (a) soil
    (b) nitrogen
    (c) grass
    (d) air

Use the following figure to answer questions 21-24.

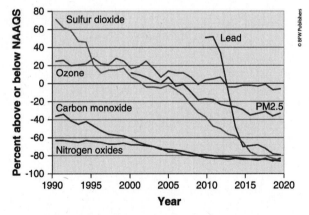

21. Which of the following legislations is most directly responsible for the trend shown in the six criteria pollutants in the graph?
    (a) The Clean Water Act
    (b) The Clean Air Act
    (c) The Kyoto Protocol
    (d) The Montreal Protocol

22. Which of the following pollutants is associated with neurological damage when it is incorporated into the bloodstream of developing infants and children?
    (a) carbon monoxide
    (b) particulate matter
    (c) sulfur dioxide
    (d) lead

23. Which of the chemicals in the graph is considered a secondary pollutant?
    (a) carbon monoxide
    (b) ozone
    (c) sulfur dioxide
    (d) particulate matter

24. Which pollutant had the smallest decrease from 2005 to 2015?
    (a) sulfur dioxide
    (b) nitrogen oxides
    (c) ozone
    (d) lead

25. Which of the following statements describes why nitrogen fixation must occur as a critical step in the nitrogen cycle?
    (a) Without nitrogen fixation, oligotrophic lakes cannot become eutrophic lakes.
    (b) Without nitrogen fixation, atmospheric nitrogen cannot be used by living organisms.
    (c) Without nitrogen fixation, elemental nitrogen cannot be returned to the atmosphere.
    (d) Without nitrogen fixation, decomposition cannot return nitrogen to the soil.

26. Which describes a difference between the process of generating electricity using coal and the process of generating electricity using nuclear energy?
    (a) Nuclear energy generates steam and coal does not.
    (b) Coal produces air pollution and nuclear energy does not.
    (c) Coal waste must be stored for millions of years and waste from nuclear energy does not.
    (d) Coal-generated electricity is much more energy efficient than nuclear-generated electricity.

27. Which chemical is most damaging to the stratospheric ozone layer?
    (a) chlorofluorocarbons
    (b) methane
    (c) sulfur dioxide
    (d) carbon dioxide

28. Which human activity increases the amount of methane released to the atmosphere?
    (a) increased rice cultivation
    (b) increased reliance on automobiles
    (c) drainage of wetland ecosystems
    (d) decreased cattle farming

Use the figure below to answer questions 29-31.

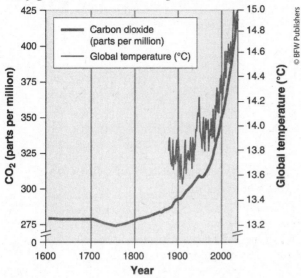

29. When carbon dioxide levels were at 300 ppm, what was the approximate global temperature in degrees Celsius?
    (a) 13.2
    (b) 14.1
    (c) 13.5
    (d) 13.8

30. What is the approximate percent change in temperature from 1900 to 2000?
    (a) 7 percent increase
    (b) 25 percent increase
    (c) 32 percent increase
    (d) 45 percent increase

31. If recent trends in global surface temperatures continues, what year do you estimate global temperature will become 15.0 Celsius?
    (a) 2000
    (b) 2100
    (c) 2500
    (d) 3000

32. Monocropping can lead to
    (a) heavy flooding.
    (b) greater susceptibility of crops to pests.
    (c) introduction of invasive species.
    (d) uncontrollable fires.

Use the following figure to answer question 33.

33. The figure shows a plate boundary. Which of the following best identifies the type of boundary and expected activity that is associated with the figure?
    (a) It is a convergent boundary where a subduction zone forms.
    (b) It is a convergent boundary where mountain ranges form.
    (c) It is a divergent boundary where seafloor spreading occurs.
    (d) It is a transform boundary where earthquakes commonly occur.

34. The net primary productivity of an ecosystem is 75 kg C/m^2/year, and producers require 20 kg C/m^2/year for their own respiration. What is the gross primary productivity of the ecosystem?
    (a) 10 kg C/m^2/year
    (b) 15 kg C/m^2/year
    (c) 50 kg C/m^2/year
    (d) 95 kg C/m^2/year

35. Which of the following describes why most of the water in the hydrologic cycle is not available for human consumption?
    (a) The majority of the water is frozen at the poles.
    (b) The majority of the water is saltwater in the oceans.
    (c) The majority of the water is found as groundwater and cannot be accessed by humans.
    (d) The majority of the water is surface water contaminated with heavy metals.

36. The greenhouse effect occurs when
    (a) gases are trapped in the stratospheric ozone layer.
    (b) infrared radiation is absorbed by gases in Earth's atmosphere.
    (c) UV light is trapped at Earth's surface.
    (d) greenhouse gases absorb UV light.

37. Which of the following characteristics describes an invasive species?
    (a) They are often found on the endangered species list.
    (b) They have narrow niches and reduced ecological tolerance.
    (c) They have limited dispersal ability without human interference.
    (d) They are often r-strategists that can reproduce quickly.

38. A scrubber on a coal burning power plant is designed to prevent the release of
    (a) sulfur dioxide.
    (b) nitrogen dioxide.
    (c) particulate matter.
    (d) carbon dioxide.

39. In one year, a population of 10,000 has 200 births, 100 deaths, 60 immigrants, and 30 emigrants. What is the population growth rate?
    (a) 1.3 percent
    (b) 9 percent
    (c) 90 percent
    (d) 2.4 percent

40. Cogeneration involves
    (a) using a fuel to generate electricity and heat.
    (b) using both coal and oil to create electricity.
    (c) substituting anthracite coal for low grade lignite coal.
    (d) operating power plants at 30 percent of maximum sustainable yield.

41. Which statement describes a small island that is far from the mainland?
    (a) It will have a high colonization rate and low extinction rate.
    (b) It will have a high colonization and high extinction rate.
    (c) It will have a low colonization rate and high extinction rate.
    (d) It will have a low colonization rate and low extinction rate.

42. The Coriolis effect is
    (a) the deflection of an object's path due to the rotation of Earth.
    (b) convection that creates air currents.
    (c) the transformation of arable land to desert because of global warming.
    (d) a change in the predator-prey relationship because of population cycles.

43. A population of 200 ducks exhibits exponential growth over two seasons and increases to 500 ducks. What is the percent change in the population of ducks?
    (a) 1.5 percent increase
    (b) 15 percent increase
    (c) 150 percent increase
    (d) 50 percent increase

44. A country with a large population that lives in extreme poverty will most likely
    (a) remain stable.
    (b) have a high infant mortality rate.
    (c) have a large environmental impact.
    (d) have a high GDP.

Read the following excerpt and answer questions 45-47.

Growing Grapes to Make Fine Wines

Wine making has its origin in the Mediterranean region, in places such as Egypt, Greece, Italy, France, and Spain. Today wine is made throughout the world, but the regions known for the finest wines are the Mediterranean, California, Chile, South Africa, and southwestern Australia. What is it about these regions that favors the production of great wines?

Wine making critically depends on having proper growing conditions for grapes. The best conditions are mild, moist winters and hot, dry summers. Mild winters are important because temperatures that fall below freezing can damage the grape vines. Hot, dry summers are important because they are moderately stressful for the plants, which causes them to create the perfect balance of sugars and acids in the grapes. The dry climate also reduces outbreaks of many grape vine diseases that are more prevalent in humid environments.

The five regions with the best growing conditions are all situated at northern and southern latitudes between 30° and 50° next to the ocean, and typically on the western side of continents. Their similarity in geographic position causes them to have air and water currents that produce comparable climates that provide similar growing conditions. Because of this, these regions also contain plants that look quite similar. Although the plant species in these five locations are not closely related, they consist of drought-tolerant grasses, wildflowers, and shrubs. In short, the plants that grow naturally in these areas, much like grapes, are well adapted to the local climate.

Because wine grapes grow best under a narrow range of climatic conditions, global climate change is causing great concern among winegrowers. Scientists recently evaluated the current and future climates in the wine region of California, where 90 percent of U.S. wine is produced. They predicted that many vineyards would have to shift northward to Oregon and Washington by 2040 because the ideal climate for wine growing will shift northward. Similarly, winemakers in France have experienced a decade of exceptionally hot and dry summers, which has made it difficult to grow the unique varieties of French wine grapes.

45. Based on the description in the article, which biome is best for growing grapes?
   (a) tropical rainforest
   (b) taiga
   (c) shrubland
   (d) desert

46. Which of the following explains why moist, humid summers impact the grape yield?
   (a) Humidity increases pathogen prevalence, which reduces the carrying capacity of grape plants.
   (b) Intense light during the summer increases photosynthesis, which increases plant growth but decreases fruit production.
   (c) Increased rainfall causes soil to become waterlogged, which decreases the sugar content of the grapes.
   (d) Increased temperature causes a decrease in plant pests, which increases the grape yield during the summer.

47. Which of the following pieces of evidence that supports climate change is affecting grape plant growth?
   (a) Grape plants must be hand pollinated because bees no longer visit the vineyards.
   (b) Grape plants are growing better in northern latitudes compared to a few decades earlier.
   (c) Grape plants have been transported from Italy to all over the world.
   (d) Grape plants require different soil nutrition than they have historically to produce the same yield.

48. Which is an example of a K-selected species?
   (a) cockroach
   (b) fish
   (c) elephant
   (d) grasshopper

49. The El Niño-Southern Oscillation would bring what type of weather conditions to southern Africa and Southeast Asia?
   (a) warmer, wetter
   (b) cooler, drier
   (c) unusually dry
   (d) unusually wet

Use the figure below to answer questions 50-51.

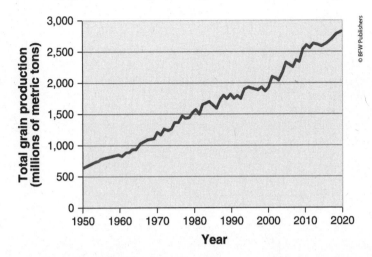

50. The figure shows Global grain production 1950–2020. Which of the following best explains the trend in grain production from 1950 to 2020?
    (a) A reduction in global rice production led to an increase in total grain production.
    (b) An increase in genetic variation in grain species allowed it to grow in a larger range across the globe.
    (c) A decrease in human population caused an abundance of grain production over time.
    (d) An increase in irrigation and fertilizer technology led to an increase in overall grain production.

51. According to the graph, which of these represents the percent change in grain production from 1980 to 2020?
    (a) 46%
    (b) 87%
    (c) -46%
    (d) -87%

Use the figure below to answer questions 52 and 53.

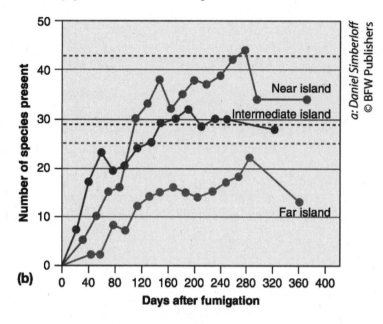

(b)

a: Daniel Simberloff
© BFW Publishers

52. Three islands of various distances from the mainland were fumigated to remove all insects. After fumigation, data was taken for a year and the number of insect species present was recorded for each island. Which statement is supported by the information in the graph?
    (a) The far island had a greater number of primary producers that supported insect species.
    (b) The intermediate island has a larger area that supports more insect species that the near and far islands.
    (c) The near island has a greater biodiversity because it is easier for insects to colonize from the mainland.
    (d) The near island has more insect species because more individuals were resistant to the initial pesticide used as a fumigant.

53. Which of the following explains the number of insect species for all three islands after day 280?
    (a) Primary succession resulted in some herbivorous insect species being replaced with other carnivorous species.
    (b) Secondary succession resulted in a loss of some insect species due to a human disturbance.
    (c) Competition for limited resources between insect species resulted in the emigration of some insects.
    (d) Competition for limited resources between insect species resulted in the local extinction of some species.

54. Which of the following is the correct order of coal type from most moisture and least heat to least moisture and most heat?
    (a) peat, lignite, bituminous, anthracite
    (b) peat, bituminous, lignite, anthracite
    (c) anthracite, bituminous, lignite, peat
    (d) bituminous, anthracite, lignite, peat

55. Which gas stays in the environment for up to 500 years and has the greatest global warming potential?
    (a) carbon dioxide
    (b) chlorofluorocarbons
    (c) methane
    (d) nitrous oxide

56. Which is an example of an anthropogenic activity?
    (a) humans burning fossil fuels to generate electricity
    (b) bees pollinating an apple tree
    (c) a volcanic eruption on a populated island
    (d) water vapor rising from a lake

57. Which tectonic location most likely contributed to the formation of the Hawaiian Islands?
    (a) transform plate boundary
    (b) subduction zone
    (c) hotspot
    (d) seafloor spreading

58. Which of the following describes how chemical weathering contributes to soil formation?
    (a) Water freezes after it infiltrates small cracks in igneous rock.
    (b) Plant roots grow into crevices of rocks and expand.
    (c) Acidic rainfall can remove potassium ions from granite to produce clay particles.
    (d) Burrowing animals can displace some of the smallest rock components.

59. Which of the following correctly identifies greenhouse gases in order of increasing global warming potential?
    (a) methane, water vapor, carbon dioxide
    (b) methane, carbon dioxide, water vapor
    (c) water vapor, carbon dioxide, methane
    (d) carbon dioxide, water vapor, methane

60. When in the sewage treatment process is large debris filtered out by screens?
    (a) primary treatment
    (b) secondary treatment
    (c) tertiary treatment
    (d) sterilization

61. A sample of radioactive waste has a half-life of 40 years and an activity level of 4 curies. After how many years will the activity level of this sample be 0.5 curies?
    (a) 60 years
    (b) 80 years
    (c) 100 years
    (d) 120 years

62. What are the two reasons for the human population's rapid growth over the past 8,000 years?
    (a) medicine and technology
    (b) smart growth and technology
    (c) organic agriculture and improved communications
    (d) lack of reliable birth control and the Green Revolution

63. Greenhouse gases in the atmosphere
    (a) prevent UV light from reaching Earth.
    (b) regulate temperatures near Earth's surface.
    (c) allow heat to be released back to space.
    (d) help Earth stay warm.

64. Which of the following explains why the Middle East has large reserves of petroleum?
    (a) Underground convection currents converge in the Middle East, bringing oil and natural gas deposits.
    (b) High levels of seismic activity in the Middle East release oil and natural gas deposits from below the sedimentary layers.
    (c) High levels of volcanic activity in the Middle East bring oil and natural gas deposits to the surface more readily.
    (d) A former ocean receded in the Middle East, leaving a desert with sedimentary layers containing oil and natural gas.

65. Which of the following is best classified as a primary air pollutant?
    (a) H_2SO_4
    (b) SO_3
    (c) H_2O_2
    (d) CO

66. Which of the following actions best addressed the problem of ozone depletion in the late 1900s?
    (a) The use of CFCs was phased out and banned in new aerosol, refrigerator, and cooling systems.
    (b) The use of radioactive fuel in nuclear power was reduced since 2000.
    (c) Catalytic converters were installed in all new vehicles manufactured after the 1980s.
    (d) Lead was banned from gasoline in the late 1900s.

67. If a material has a radioactivity level of 100 curies and has a half-life of 50 years, how many half-lives will have occurred after 100 years?
    (a) 1
    (b) 2
    (c) 10
    (d) 25

68. Thick curtains on windows can be an aspect of
    (a) active solar design.
    (b) energy star technology.
    (c) passive solar design.
    (d) a tiered rate system.

Use the following figure to answer question 69.

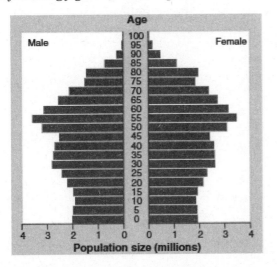

69. The figure shows an age structure diagram for Germany. Which statement explains why Germany has a declining population size?
    (a) Germany is a developed nation with the majority of its adults beyond the child-bearing age.
    (b) Germany is a developing nation with low access to healthcare.
    (c) The majority of older Germans are migrating out of the country.
    (d) There is an imbalance in the male: female sex ratio.

Use the following figure to answer question 70.

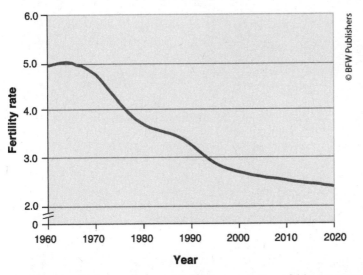

70. The graph shows the total fertility rate of world human population over time. Based on the graph which statement is supported?
    (a) The human population is decreasing in size because the fertility rate is decreasing.
    (b) The human population size is stable because the fertility rate is approaching two.
    (c) The human population size is still growing because the total fertility rate is above the replacement-level fertility of 2.1.
    (d) The human population is decreasing because humans have exceeded their carrying capacity.

71. Which of the following best describes the Ogallala aquifer over the past fifty years?
    (a) The water level of the Ogallala aquifer has decreased due to increased grazing of livestock in the central United States.
    (b) The water level of the Ogallala aquifer has decreased due to increased irrigation of agricultural land in the central United States.
    (c) The water level of the Ogallala aquifer has increased due to decreased urbanization of land in the central United States.
    (d) The water of the Ogallala aquifer has been contaminated by saltwater intrusion in the central United States.

72. If a country's population growth rate is 7 percent, what is the country's doubling time?
    (a) 5 years
    (b) 10 years
    (c) 42 years
    (d) 72 years

Use the following graph to answer question 73.

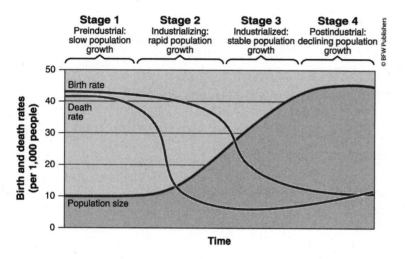

73. In the graph, at which stage does the birth rate fall below the death rate?
    (a) stage 1
    (b) stages 1 and 2
    (c) stage 3
    (d) stage 4

74. In the graph, at which stage is the population size relatively stable?
    (a) stage 1 only
    (b) stage 2 only
    (c) stage 4 only
    (d) stages 1 and 4

75. Why do levels of carbon dioxide in the atmosphere vary with the seasons?
    (a) More fossil fuels are burned in the winter for heat.
    (b) More fossil fuels are burned in the summer for cooling.
    (c) Livestock production increases each spring.
    (d) Deciduous trees do not take in carbon dioxide in the fall and winter.

76. Which is a major environmental impact associated with deforestation?
    (a) removal of soil nutrients
    (b) acid rain
    (c) depletion of the stratospheric ozone layer
    (d) invasive species taking over the deforested area

77. Which of the following practices would reduce the entanglement and intestinal blockage that affects certain marine birds and sea turtles?
    (a) a reduction in endocrine disruptors used as herbicides
    (b) a reduction in the nitrogen and phosphate levels used in synthetic fertilizers
    (c) a reduction in plastic debris that is disposed of in the ocean
    (d) a reduction in the transport of oil in large shipping vessels

78. Which of the following has the potential to convert an oligotrophic lake into a eutrophic lake?
    (a) an influx of nitrates from agricultural runoff
    (b) an influx of pesticides used to control insects from agricultural runoff
    (c) an increase in oil spills and leaks that occurs from the drilling and transport of petroleum
    (d) a decrease in the lake pH from acid deposition

79. The process of recycling an aluminum can into a new aluminum can is an example of
    (a) closed-loop recycling.
    (b) life-cycle analysis.
    (c) restoration.
    (d) open-loop recycling.

80. Which describes the results of an experiment that determined the LD50 of a chemical?
    (a) 2 out of every 100 rats died.
    (b) 25 out of every 50 rats died.
    (c) 1 out of every 50 rats got sick.
    (d) 50 out of every 100 rats got sick.

SECTION II: Free-Response Questions

Write your answer to each part clearly. Support your answers with relevant information and examples. Where calculations are required, show your work. Use a separate sheet of paper if necessary.

1. The town of Fremont is concerned about water pollution because of sewage odor coming out of the local river. The town council has appointed a task force to investigate this problem. The researchers on the task force take water samples from four sites along the river. They test for nitrate, phosphate, and coliform levels at each of the sites. The results are shown in the table below.

Site	Nitrate level	Phosphate level	Coliform level
I	0.51 mg/L	0.1 mg/L	<25 per 100 ml
II	0.48 mg/L	0.12 mg/L	<25 per 100 ml
III	4.2 mg/L	1.8 mg/L	>2500 per 100 ml
IV	1.2 mg/L	0.45 mg/L	>2500 per 100 ml

(a) **Identify** one independent variable and one dependent variable in the research team's experimental design. **Identify** the sample site that indicates that a point source is affecting the river.

(b) The research team identifies an output pipe from a sewage treatment plant that is contributing to the changes in the river.

  i. **Identify** the stage of wastewater treatment that removes phosphate and nitrates from the water.

  ii. **Describe** a process during sewage treatment that would reduce the number of coliform bacteria in the water.

  iii. **Identify** a federal regulation that would monitor the quality of the river water.

(c) Based on the information in the data table, **predict** the changes in dissolved oxygen that would be observed at sample sites II and III. **Explain** why these changes would be observed.

(d) The task force also wants to investigate the potential for nonpoint source pollution to affect the river.

  i. **Describe** a nonpoint source of pollution that could affect the river.

  ii. **Propose** one solution to the source you identified in part (i) to reduce the effects on the river.

2. Many scientists are concerned about anthropogenic activities that they believe have contributed to global climate change.

(a) **Identify** two atmospheric gases with a global warming potential of greater than 1.

(b) **Describe** the trend shown in the graph below as it relates to carbon dioxide level and global temperature. **Describe** two human activities that could reduce the trend shown in the graph.

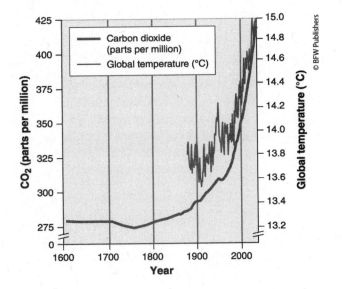

(c) **Explain** how the trend in the graph relates to Arctic ice levels. **Describe** the type of feedback loop involved in this process.

(d) **Explain** how the trend shown in the graph impacts the pH of the ocean. **Describe** one impact on aquatic ecosystems that could occur due to this trend.

(e) Identify one natural process in the carbon cycle that contributes to the carbon dioxide levels shown in the graph.

3. A high school in Florida has decided to install photovoltaic panels on its roof. The cost to install the panels will be $15,000. Currently, the school pays $0.10 per kWh and the average monthly electricity use for the school is 20,000 kWh. The panels will produce 2,000 kWh per month.

(a) **Calculate** how much the school paid for electricity during a 30-day month before the installation of the photovoltaic panels. **Show** your work.

(b) **Calculate** how many years will it take for the school to recoup the cost of its purchase. **Show** your work.

(c) If the school decides to invest $30,000 more in solar panels, calculate how much more energy they could produce. (Express your answer in kWh.) **Show** your work.

(d) **Describe** two practices the school could implement to lower its energy use.

(e) The school decides to incorporate passive solar design as a way to decrease its energy consumption. **Describe** two passive solar energy techniques the school could implement.